Mastering Information Security Compliance Management

A comprehensive handbook on ISO/IEC 27001:2022 compliance

Adarsh Nair

Greeshma M. R.

BIRMINGHAM—MUMBAI

Mastering Information Security Compliance Management

Group Product Manager: Pavan Ramchandani
Publishing Product Manager: Prachi Rana
Senior Content Development Editor: Adrija Mitra
Technical Editor: Irfa Ansari
Copy Editor: Safis Editing
Project Coordinator: Manisha Singh
Proofreader: Safis Editing
Indexer: Manju Arasan
Production Designer: Jyoti Chauhan
Marketing Coordinator: Rohan Dobhal

First published: August 2023

Production reference: 1140723

Published by Packt Publishing Ltd.
Livery Place
35 Livery Street
Birmingham
B3 2PB, UK.

ISBN 978-1-80323-117-4

www.packtpub.com

Contributors

About the authors

Adarsh Nair is the global head of information security at UST. He is a recognized information security strategist, author, and keynote speaker. Adarsh holds the title of **Fellow of Information Privacy (FIP)** by IAPP and is a Google Hall of Fame honoree. He serves as co-chair of OWASP Kerala Chapter, an IAPP exam development board member, and an EC-Council advisory board member.

With a decade of experience, Adarsh specializes in information security governance, risk and compliance, business continuity, data privacy, ethical hacking, and threat identification and mitigation. He maintains expertise through memberships, training, and certifications, including CISSP, CIPM, CIPP/E, LPT, OSCP, and ISO Lead Auditor.

Adarsh has authored two books, published numerous articles and research papers, and delivered impactful presentations at national and international conferences, establishing himself as a thought leader in information security.

Greeshma M. R. is an entrepreneur and seasoned freelance technology writer, specializing in technology domains, especially information security and Web 3.0. She is interested in exploring the intersection of technology and humanity, as well as the social aspects of technology. Her areas of interest also encompass innovation, sustainable development, gender, and society. She is a co-author and publisher of two books and holds a certification as an ISO 27001 Lead Auditor.

Having worked in the IT and knowledge and innovation management domains, Greeshma possesses an interdisciplinary perspective that enriches her approach. She has actively contributed to establishing an innovation ecosystem among students via communities of practice, fostering a culture of creativity and collaboration.

About the reviewers

Fernando Rose is a highly experienced professional in the security industry with 30+ years of experience. He started as a customer support provider at Motorola, where he worked for six years. In 2003, he founded FAR Solutions Limited, combining his Lead Auditor qualifications with critical InfoSec skills. Fernando has co-authored a book and holds several certifications, including Lead SIA Auditor and Lead Auditor. He audits and supports various sectors, including forensics, defense, clinical, commercial, and telecoms, ensuring their safety and integrity.

I thank my partner, mentors, and sturdy reference books for supporting my InfoSec career. Trust is essential, and ISO 27001 forms its basis. To succeed, blend technical skills with interpersonal abilities. I am incredibly grateful for the guidance that has helped me contribute to information security.

Krutika Vadlakonda has a master's in information systems and cybersecurity from Georgia State University and has worked in the field of information technology and cybersecurity for over 10 years. She started her career in a Big Four audit firm, where she consulted enterprises and conducted audits to assess software license compliance. She has certified companies on ISO 27001 as a Lead Auditor and led SOC 2 attestation reviews for Fortune 500 companies. Currently, Krutika works for a leading cloud service provider and manages cybersecurity programs to keep customers secure at scale. She also consults leading global enterprises, helping them with security strategy and building robust security governance programs in the cloud.

I truly believe that security compliance programs are the binding glue to keep security risks low for organizations. You can learn about (almost) all the core areas in the cybersecurity domain by learning about a compliance framework such as ISO 27001. I am grateful to my managers/mentors for supporting me in my cybersecurity journey, and for all the support and encouragement from my family, who stand by me when days are tough.

Table of Contents

Part 1: Setting the Stage – Definitions, Concepts, Principles, Standards, and Certifications

1

Foundations, Standards, and Principles of Information Security   3

Part 2: The Protection Strategy – ISO/IEC 27001/02 Design and Implementation

5

ISMS – Phases of Implementation   73

6

Information Security Incident Management   91

7

Case Studies – Certification, SoA, and Incident Management   105

Part 3: How to Sustain – Monitoring and Measurement

8

Audit Principles, Concepts, and Planning   143

9

Performing an Audit   165

10

Audit Reporting, Follow-Up, and Strategies for Continual Improvement   175

11

Auditor Competence and Evaluation   185

12

Case Studies – Audit Planning, Reporting Nonconformities, and Audit Reporting   193

Appendix – Terms and Definitions   201

Index   207

Other Books You May Enjoy   216

Preface

In the rapidly expanding digital age, data has gained the moniker of the "new oil," highlighting its immense significance. Consequently, the security and management of this invaluable resource have emerged as a paramount concern. In response, international standards have been established to guide organizations in implementing and maintaining robust **Information Security Management Systems (ISMSs)**. *Mastering Information Security Compliance Management*, offers an in-depth, comprehensive exploration of these standards, specifically ISO/IEC 27001 and 27002.

From foundational principles to intricate processes, this book covers the entire spectrum of information security through 12 detailed chapters. Beginning with a broad overview of information security and the role of standards, it then delves into the specifics of ISO 27001 and its applications. It discusses the implementation of an ISMS, provides insight into the intricate details of ISO 27001 and 27002 control references, and navigates the crucial stages of risk assessment and management. Moreover, it illuminates the complexities of developing an ISMS tailored to unique business contexts and tackles the crucial aspect of information security incident management.

You will be guided through a series of real-life case studies highlighting the practical application of the concepts discussed, along with a thorough examination of audit principles, planning, performance, and reporting. The final chapters explore strategies for continual improvement of an ISMS, the evaluation of auditor competence, and the ethics of the auditing profession.

The goal of this handbook is to equip you with a nuanced understanding of ISO/IEC 27001/27002 standards, enabling you to effectively implement, audit, and enhance an ISMS in your organization, ensuring data security, regulatory compliance, and overall organizational resilience. This book is an essential resource for all professionals engaged in the world of information security.

Who this book is for

This book is designed for a diverse readership looking to enhance their understanding and application of ISO/IEC 27001/27002 standards. It is especially valuable for information security professionals, including information security managers, compliance officers, and IT managers, who are responsible for implementing, managing, and auditing an ISMS. Consultants who assist organizations in establishing an ISMS will also find this book highly beneficial. Furthermore, executives and decision-makers aiming to understand the relevance and benefits of implementing ISO/IEC 27001/27002 in their organization can leverage this resource. Academics and students in fields such as information technology, business administration, and cybersecurity may also find this handbook helpful in their studies and research. In essence, this book is a crucial companion for anyone seeking to understand, implement, manage, or audit ISO/IEC 27001/27002 standards in the pursuit of robust information security.

What this book covers

In *Mastering Information Security Compliance Management: A comprehensive handbook on ISO/IEC 27001:2022 compliance*, each chapter contributes to building a holistic understanding of the ISO/IEC 27001/27002 standards and their implementation.

Chapter 1, Foundations, Standards, and Principles of Information Security, establishes the groundwork, explaining the core principles of information security and the role of ISO/IEC 27000 standards, specifically ISO/IEC 27001, to develop a robust ISMS.

Chapter 2, Introduction to ISO 27001, provides an in-depth exploration of ISO 27001, its operational model, the benefits, and the processes involved in achieving accreditation from recognized bodies.

Chapter 3, ISMS Controls, focuses on the controls outlined in ISO 27001/27002, detailing their interpretation and application based on the specific business context.

Chapter 4, Risk Management, dives into the integral components of the ISO 27001 framework, emphasizing the role of risk assessment, management, and the necessity of a risk register.

Chapter 5, ISMS – Phases of Implementation, takes you through the various stages involved in developing an ISMS, illustrating how to tailor control implementation to the specific context of a business.

Chapter 6, Information Security Incident Management, covers the essential aspects of incident management, highlighting the importance of comprehensive incident management plans.

Chapter 7, Case Studies – Certification, SoA, and Incident Management, offers practical insights through real-world case studies, focusing on certification, the **Statement of Applicability** (**SoA**), and incident management.

Chapter 8, Audit Principles, Concepts, and Planning, delves into the principles of auditing, introducing different types of audits and outlining the processes involved in planning for audits.

Chapter 9, Performing an Audit, guides you through the audit process, from data collection and system effectiveness assessment to the formulation of reports and recommendations.

Chapter 10, Audit Reporting, Follow-Up, and Strategies for Continual Improvement, discusses the importance of audit reporting, follow-up processes, and strategies for the continual improvement of an ISMS.

Chapter 11, Auditor Competence and Evaluation, focuses on the competencies, responsibilities, and ethical conduct required of auditors in the auditing process.

Chapter 12, Case Studies – Audit Planning, Reporting Nonconformities, and Audit Reporting, concludes the book with practical examples and real-world scenarios, focusing on audit planning, reporting nonconformities, and audit reporting.

The entire book offers a comprehensive understanding of the ISO/IEC 27001/27002 standards, presenting both theoretical knowledge and practical application, aiding you in implementing, auditing, and enhancing an ISMS in your organization.

Conventions used

There are a few text conventions used throughout this book.

Bold: Indicates a new term, an important word, or words that you see on screen. For instance, words in menus or dialog boxes appear in **bold**. Here is an example: "**ISO 27035** is the standard that talks in detail about information security incident management. Information security incidents and vulnerabilities can be identified, documented, assessed, responded to, managed, and used to drive future efforts to strengthen security."

Italics: Highlights important parts of a sentence and is also used when referring to another chapter, an image or table, or a section of the same chapter. Here is an example: "There are three different aspects of auditor competence that are identified in the *ISO 19011* standard for management system auditing – *personal behavior*, *technical competence*, and *auditing competence*."

> **Tips or important notes**
> Appear like this.

Get in touch

Feedback from our readers is always welcome.

General feedback: If you have questions about any aspect of this book, email us at customercare@packtpub.com and mention the book title in the subject of your message.

Errata: Although we have taken every care to ensure the accuracy of our content, mistakes do happen. If you have found a mistake in this book, we would be grateful if you would report this to us. Please visit www.packtpub.com/support/errata and fill in the form.

Piracy: If you come across any illegal copies of our works in any form on the internet, we would be grateful if you would provide us with the location address or website name. Please contact us at copyright@packtpub.com with a link to the material.

If you are interested in becoming an author: If there is a topic that you have expertise in and you are interested in either writing or contributing to a book, please visit authors.packtpub.com.

Share Your Thoughts

Once you've read *Mastering Information Security Compliance Management,* we'd love to hear your thoughts! Scan the QR code below to go straight to the Amazon review page for this book and share your feedback.

https://packt.link/r/1803231173

Your review is important to us and the tech community and will help us make sure we're delivering excellent quality content.

Download a free PDF copy of this book

Thanks for purchasing this book!

Do you like to read on the go but are unable to carry your print books everywhere?

Is your eBook purchase not compatible with the device of your choice?

Don't worry, now with every Packt book you get a DRM-free PDF version of that book at no cost.

Read anywhere, any place, on any device. Search, copy, and paste code from your favorite technical books directly into your application.

The perks don't stop there, you can get exclusive access to discounts, newsletters, and great free content in your inbox daily

Follow these simple steps to get the benefits:

1. Scan the QR code or visit the link below

https://packt.link/free-ebook/9781803231174

2. Submit your proof of purchase
3. That's it! We'll send your free PDF and other benefits to your email directly

Part 1: Setting the Stage – Definitions, Concepts, Principles, Standards, and Certifications

Part 1, encompassing *Chapter 1* and *Chapter 2*, is the cornerstone of this book, setting the scene with an exploration of information security's fundamental principles and the ISO 27001 standard. *Chapter 1* explains the basics of information security – confidentiality, integrity, and availability – and introduces the ISMS framework. *Chapter 2* builds on this foundation by examining the PDCA process model integral to ISO 27001, providing a SWOT analysis of ISMS implementation, and underscoring the importance of accreditations and certifications. This section lays a robust groundwork for a comprehensive understanding of the ISO/IEC 27001/27002 standards.

This part has the following chapters:

- *Chapter 1, Foundations, Standards, and Principles of Information Security*
- *Chapter 2, Introduction to ISO 27001*

1
Foundations, Standards, and Principles of Information Security

In today's information-centric environment, the concept of information security is paramount and is now on par with other business functions. Irrespective of their market share, private or public status, or geographical location, businesses are being pushed to move online in order to stay relevant.

In the 21st century, we have all experienced the information revolution. Data is stimulating the information revolution in the same way that oil catalyzed the industrial revolution. In today's environment, data is the raw resource that must be studied, interpreted, and retrieved with care in order to provide significant insights to its users.

The difference between oil and data is that the volume of oil is reducing across the world, whereas the amount of data is growing day by day. Data has become a valuable commodity and fuel source in today's world.

On the other hand, data-related cybercrime such as data theft is expanding exponentially. A data breach occurs when a company unwittingly exposes critical information that might cause damage to a company's reputation, brand value, and customer trust, or even result in regulatory penalties.

The average cost of a data breach was $4.35 million in the year 2022, according to IBM's Cost of a Data Breach Report 2022. While the average cost per record was $164 in 2022, the cost per record has climbed considerably since 2020. Hackers are primarily interested in a company's customer information because they can use it to blackmail the company or sell the information to competitors. Data has become, on average, more valuable than any other asset. Information security principles guide the entire concept of data security.

This chapter will explain the fundamentals of **Information Security**, including why it's important and how security frameworks can help reduce risk and develop a mechanism to manage information security across an enterprise. The key topics covered are the following:

- The CIA triad
- Information security standards
- Using an information security management system
- The ISO 27000 series

The CIA triad

InfoSec, the shorthand for information security, refers to procedures designed to secure data from unauthorized access or modification, even when the data is at rest or in transit. It covers a broad range of topics, including safeguarding your digital assets, which is where you hold sensitive data.

Information security relies on three pillars known as the **CIA Triad**: Confidentiality, Integrity, and Availability, the preservation of which is defined in ISO/IEC 27000. See *Figure 1.1* for a visual representation of the following three pillars:

- **Confidentiality** – Providing access only to authorized personnel who need access
- **Integrity** – Maintaining the information's accuracy and completeness
- **Availability** – Making sure the information is available to authorized users when they need it

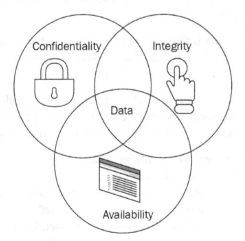

Figure 1.1 – CIA triad

Let's see what each of the pillars in the triad means for information security.

Confidentiality

When an organization takes steps to keep its information private or secret, it is referred to as confidentiality. In the real world, this means limiting who has access to data in order to keep it safe from unwanted disclosure. Unauthorized disclosure of information or unauthorized access to information systems can be prevented by implementing confidentiality safeguards. For the confidentiality principle to be effective, sensitive information must be protected and only those who need access to accomplish their job responsibilities should be able to see or access it.

Confidentiality is required to prevent sensitive information from leaking to the wrong people. It is possible to safeguard user data by using authentication controls such as **passwords** and the encryption of data that is in transit or at rest to keep it confidential.

Integrity

Integrity refers to the ability of a person or thing to stand on their or its own two feet. In the same sense, integrity in information security entails the safeguarding of data from uncontrolled or unauthorized additions, deletions, or modifications. Integrity is based on the idea that data can be trusted to be accurate and not improperly altered.

The idea of non-repudiation, or the inability to refute anything, is closely linked to integrity. Non-repudiation of information and services is ensured by this criterion and thus provides traceability of the actions conducted on them. At all times, accuracy and consistency in data are vital. You must be prepared to show that document credibility has been maintained, particularly in legal circumstances, when it comes to integrity. Hashing, digital signatures, and digital certificates are often employed to ensure the integrity of data.

Availability

It is useless for a business to have valuable systems, apps, or data that can't be easily accessed by the people who need them. Being available implies all systems and apps are working as expected, and resources are available to authorized users in a timely and reliable manner. The goal of availability is to ensure that data and services are available when needed to make decisions.

The accessibility of the system and services provided to authorized users is dependent on the availability factor because the system and services should be available whenever the user needs them. Redundancy of important systems, hardware fault tolerance, frequent backups, extensive disaster recovery plans, and so on, are all ways to assure availability.

Accountability and cyber resilience

Accountability entails assigning explicit obligations for information assurance to each person who interacts with an information system. A manager responsible for information assurance can readily quantify the responsibilities of an employee within the context of the organization's overall information security plan. A policy statement saying that no employee shall install third-party software on company-owned information infrastructure is one example. To be resilient in the face of cyberattacks, a business must be capable of anticipating them, preparing for them, and responding to them appropriately. This aids an organization in combating cyber threats, reducing the severity of attacks, and guaranteeing that the company continues to exist even after an attack has taken place. This is cyber resilience.

The CIA triad forms the foundation of information security standards such as ISO/IEC 27001. Let's now look at some of the standards that are accessible in the information security sector.

Information security standards

Standards provide us with a common set of reference points that allow us to evaluate whether an organization has processes, procedures, and other controls that fulfill an agreed-upon minimum requirement. Depending on the needs of the business or stakeholders, an organization may build and manage its own procedures in accordance with information security principles. It offers third parties such as **customers**, **suppliers**, and **partners** confidence in an organization's capacity to deliver to a specific standard if that business is compliant with the standard.

This can also be a marketing strategy whereby the company can gain a competitive advantage over other organizations. When customers are evaluating a company's products or services, for example, an organization that is compliant with a security standard may have the edge over a competitor who is not.

On the other hand, some regulatory and legal requirements may specify certain standards that must be met in certain circumstances. Suppose your company stores, processes, or transmits cardholder data. In this case, you must be in compliance with the **Payment Card Industry Data Security Standard** (**PCI DSS**). There are a variety of organizations involved in accepting credit and debit cards and the PCI DSS applies to each and every one of them. Major credit card firms such as Visa and Mastercard have identified these criteria as being the industry benchmark. Failure to comply with these standards may result in fines, increased processing fees, or even the refusal to do business with certain credit card companies.

Furthermore, if you are supposed to be compliant with a standard but are not, and you suffer a security breach as a result, you may be subject to legal action from the consumers who were harmed as a result of the breach.

Standards can also assist firms in meeting regulatory requirements such as those imposed by the **Data Protection Act, Sarbanes–Oxley Act (SOX), Health Insurance Portability and Accountability Act (HIPAA)**, and other similar legislation. Utilizing standards to establish a solid foundation for managing and protecting your information systems will make it easier for your organization to comply with current and future regulatory obligations than for an organization that does not use standards.

Let's have a quick look at some of the important standards in the field of information security.

The ISO/IEC 27000 family of information security standards

The **ISO 27000 Family of Information Security Management Standards** is a collection of security standards that form the basis of best-practice information security management. **ISO 27001**, which establishes the requirements for an **Information Security Management System (ISMS)**, is the series' backbone.

ISO 27001 is a global standard that defines the criteria for an ISMS. The structure of the standard is intended to assist companies in managing their security procedures in a centralized, uniform, and cost-effective manner.

Payment Card Industry Data Security Standard (PCI DSS)

The **PCI Security Standards Council (PCI SSC)** is an independent organization founded by Visa, MasterCard, American Express, Discover, and JCB to administer and oversee the PCI DSS. According to this regulation, companies, financial institutions, and merchants must comply with a set of security criteria when dealing with cardholder data. A secure environment needs to be maintained to receive, process, store, and transmit cardholder information.

Federal Information Security Management Act (FISMA)

The **Federal Information Security Management Act (FISMA)** is a set of data security principles that federal agencies must follow in order to preserve and secure their data. Private enterprises that have a contractual connection with the government are likewise subject to FISMA's regulations.

Government data and information are protected, and governmental expenditure on security is kept under control. FISMA established a set of regulations and standards for government institutions to follow in order to meet data security objectives.

Health Insurance Portability and Accountability Act (HIPAA)

In order to protect the privacy and confidentiality of patient health information, the **Health Insurance Portability and Accountability Act (HIPAA)** of 1996 mandated the development of national standards. This is also known as the **Kennedy–Kassebaum Act**.

Health information that may be used to identify a specific individual is covered by the HIPAA, which applies to all forms of **protected health information** (**PHI**). All covered entities such as healthcare providers, health plans, and healthcare clearinghouses are under the Health Insurance Portability and Accountability Act of 1996.

Due to the security standard in place, patients may rest easy knowing that the fundamental health-related information they provide will be kept confidential.

NIST Cybersecurity Framework (NIST CSF)

The NIST framework for cybersecurity is a useful tool for organizing and improving your cybersecurity program. In order to assist businesses to establish and enhance their cybersecurity posture, this set of best practices and standards was put together.

A cybersecurity program built on the **NIST Cybersecurity Framework** (**NIST CSF**) is widely regarded as the industry standard. To assist enterprises in managing and reducing cybersecurity risk, the NIST CSF provides suggestions based on existing standards, guidelines, and practices.

No matter where they are located, all organizations may use this framework despite its original intent to protect important US infrastructure corporations.

SOC reporting

An internal control report developed by the **American Institute of Certified Public Accountants** (**AICPA**) is called the **System and Organization Controls** (**SOC**) for service organizations. Using SOC reports, service providers may increase their customers' trust in the services they deliver, as well as their own internal control over those services. SOC 1, SOC 2, and SOC 3 are the three types of reports that can be used based on the requirements.

The **SOC 1: SOC for Service Organization: ICFR** report (type 1 or 2) evaluates an organization's internal financial reporting controls in order to evaluate the impact of the controls of the service organization on the financial statements of its customers.

The purpose of the **SOC 2: SOC for Service Organizations: Trust Services Criteria** report (type 1 or 2) is to reassure customers, management, and other stakeholders about the appropriateness and efficacy of the service organization's security, availability, processing integrity, confidentiality, and privacy measures (trust principles).

The **SOC 3: SOC for Service Organizations: Trust Services Criteria for General Use** report is a condensed version of the SOC 2 (type 2) report for consumers who want assurance regarding the security, availability, processing integrity, confidentiality, or privacy controls of service organizations. SOC 3 reports may be freely disseminated since they are general-purpose reports.

Cybersecurity Maturity Model Certification (CMMC)

To examine its contractors' and subcontractors' security, competence, and resilience, the US Department of Defense uses the **Cybersecurity Maturity Model Certification (CMMC)**. This framework's goal is to make the supply chain more secure by eliminating vulnerabilities. Control practices, security domains, procedures, and capabilities make up the CMMC.

Five levels of management are utilized in the CMMC architecture. The lowest maturity level is level 1, while the highest is 5. There are tiers of service that contractors are expected to provide depending on the amount of data they manage under the contract. Achieving each level of certification necessitates meeting particular standards by collaborating with various cybersecurity elements.

Information security standards help prove that the organization meets the stipulated data security levels and is compliant. These standards need to be effectively implemented and managed, and that is the role of an **Information Security Management System (ISMS)**.

Using an information security management system

It is an open secret that every business is a target for cyberattacks. Despite the fact that data breaches are growing increasingly catastrophic, many firms still believe they will never be victims. If you have strong defenses, you can prevent most attacks and prepare for a breach. People, procedures, and technology are the three ISMS pillars that help an organization to achieve adequate security compliance.

An ISMS demonstrates the organization's approach to information security. It will help you detect and respond to threats and opportunities posed by your sensitive data and any associated assets. This safeguards your organization and business processes from security breaches and protects them from disruption if they occur.

An ISMS is a framework for establishing, monitoring, reviewing, maintaining, and enhancing an organization's information security compliance in order to achieve business and regulatory requirements. It is designed to identify, mitigate, and manage risks effectively by conducting a risk assessment and considering the firm's risk appetite. Analyzing information asset protection requirements and implementing appropriate controls to ensure that these information assets are protected, as needed, helps in the effective deployment of an ISMS. An ISMS consists of the policies, processes, guidelines, allocated resources, and associated activities that an organization controls together to protect its information assets.

Information is data that is organized and processed, and which has a meaning in context for the receiver. Like other key business assets, it is critical to the operation of an organization and, as such, must be adequately secured. Electronic or optical media may store digital information (such as data files), while paper-based information (such as documents) or tacit knowledge among personnel can be used to store information as well. It can be sent via courier, email, or verbal conversation, among other methods. It must be protected regardless of how it is sent.

Information is reliant on information and communications technologies and infrastructure in many enterprises. This technology is frequently a critical component of an organization, assisting in generating, processing, storing, transferring, protecting, and destroying information.

Confidentiality, availability, and integrity form the three main dimensions of information security. Implementing and managing adequate security controls as part of an ISMS that addresses a wide range of possible risks helps reduce the effect of information security events, thereby ensuring long-term organizational success and continuity.

Controls are implemented according to the risk management process and managed through an ISMS to safeguard identified information assets in order to accomplish information security. These controls include policies and processes, as well as procedures and organizational structures. In order to meet the organization's specific information security and business objectives, controls must be established, implemented, evaluated, reviewed, and, if necessary, upgraded. A company's business activities must be taken into consideration while implementing information security controls.

Management entails actions aimed at directing, controlling, and continuously improving an organization within proper organizational structures. Management activities are the actions, styles, or practices of organizing, managing, directing, controlling, and regulating resources. Small enterprises may have a flat management structure with just one person, whereas large corporations may have hierarchies with dozens or even hundreds of people.

From an ISMS perspective, management includes the oversight, support, and decision-making essential to meet the business objectives and regulatory requirements by ensuring the security of the organization's information assets. Information security management is exemplified by developing and implementing necessary policies, processes, and guidelines, which are subsequently implemented across the organization.

A management system makes use of a framework to help an organization accomplish its goals. Incorporating a management system means considering the organization's structure, policies, and planning activities, along with roles and duties.

An information security management system helps an organization to do the following:

- Meet all interested parties' information security requirements
- Design and execute the organization's tasks more effectively
- Realize the information security goals
- Comply with all applicable laws, regulations, and industry best practices
- Ensure systematic management of information assets

Principle of least privilege and need to know

According to the **Principle of Least Privilege** (**POLP**), a person should only be granted the privileges necessary to carry out their job. POLP also limits who has access to apps, systems, and processes to only those who are authorized. POLP is implemented in the **Role-Based Access Control** (**RBAC**) system, which guarantees that only information relevant to the user's role is accessible and prohibits them from obtaining information that is not relevant to their role.

Following the POLP lowers the danger of an attacker compromising a low-level user account, device, or application, giving them access to vital systems or sensitive data. By using the POLP, compromises can be contained to the source location, rather than spreading throughout the entire system.

The need-to-know concept can be enforced through user access controls and permission procedures, and its goal is to ensure that only individuals who are authorized have access to the information or systems they need to perform their jobs.

According to this rule, a user should only have access to the data necessary to perform their work. Need to know implies that access is granted based on a legitimate requirement and is then revoked at the end of the project.

An ISMS reflects an organization's attitude toward protecting data. Implementing an ISMS can be particularly important to an organization in protecting its own data as well as its clients'.

Why is an ISMS important?

An ISMS is crucial because they provide a structure for safeguarding a company's most confidential data and assets. They aid businesses in spotting threats to their data and assets and devising strategies to counteract them.

According to recent PwC research, one in every four businesses worldwide has had a data breach that cost them between $1 and $20 million or more in the last three years. The average cost of a data breach in 2022 was $4.35 million, according to IBM and Ponemon's 2022 research. Last year, the average breach cost $4.24 million. From $3.86 million in 2020, the average cost has increased by 12.7%.

A leading e-commerce company was fined $877 million for breaking GDPR cookie regulations, a telecom company paid $350 million to resolve a class action lawsuit over a data breach in early 2021, and a software company was penalized $60 million for misleading Australian customers about location data.

A study by the **British Standards Institution** (**BSI**) found that 51.6% of organizations with a certified ISMS reported fewer security incidents.

An ISMS helps an organization devise a plan for handling sensitive information, such as personal and confidential business information, in a systematic way. This reduces the chances of a data breach and the financial and reputational damage it can cause. An ISMS helps businesses comply with applicable laws and regulations, such as the GDPR and HIPAA, in order to avoid penalties and reputational damage.

It is necessary to address the risks connected with an organization's information assets. All of an organization's information assets have an associated risk, which needs to be addressed through risk management. Information security needs risk management, which incorporates risks posed by physical, human, and technological threats to all types of information stored or used by the company. This strategic choice must be seamlessly integrated, scaled, and updated to match the organization's needs when an ISMS is designed for an organization.

The design and execution of an ISMS are influenced by a variety of factors, including the organization's goals, security requirements, business processes, and size and structure. All stakeholders in the firm, including consumers, suppliers, business partners, shareholders, and other key third parties, must be taken into account while designing and operating an ISMS.

The importance of an ISMS cannot be overstated. An ISMS is a key facilitator of risk management initiatives in any sector. Data access and management become more challenging to govern due to public and private network interconnectivity and the sharing of information assets. Additionally, the proliferation of mobile storage devices carrying information assets has the potential to erode the effectiveness of existing controls.

Businesses that adhere to the ISMS family of standards show their ability to adopt consistent and mutually acknowledged information security principles to their clients and partners. The design and development of information systems do not always take information security requirements into account. The level of information security compliance that may be accomplished using technological approaches is restricted. It may be ineffective unless complemented by appropriate management and policy/procedures within the context of an ISMS.

It can be difficult and expensive to integrate security into a fully operational information system. An ISMS requires careful preparation and attention to detail because it entails establishing which controls are in place. As an example, in order to provide appropriate permission and access limitation to information assets or a facility, access controls need to be designed and put into place. The controls may be technological, physical, administrative, or a combination of all three, depending on the nature of the business and its information security needs.

Companies can have more confidence in the security of their information assets due to the effective deployment of an ISMS, which helps them identify and analyze risks, implement appropriate controls, and meet regulatory requirements.

In conclusion, an ISMS is valuable because it assists businesses in safeguarding private data and assets, mitigating the financial impact of data breaches, and meeting regulatory requirements. Using an ISMS enables organizations to manage their own data assets and those entrusted to them by third parties. Let's look at the ingredients that make an ISMS implementation successful.

Key factors of an effective ISMS

Several factors contribute to the effectiveness of an ISMS implementation that helps a company to achieve its business goals. The following are the most important criteria for success:

- Documented information on information security goals, policies, procedures, and implementations that are available and in alignment with the business objectives of the organization.

- Architecture, implementation, tracking, maintenance, and enhancement of the information security framework in accordance with the organization's culture and values.

- All levels of management, especially senior management, showing their full support and commitment. The implementation should start from the top leadership to bring the right culture throughout the ISMS processes. This is known as a top-down approach.

- Risk management and information security needs are clearly understood.

- Successful implementation of information security awareness, training, and education programs that inform all interested parties, including employees, about the defined information security obligations of the organization and motivates them to abide by them.

- An effective process for managing information security incidents.

- An effective strategy and process for ensuring business continuity.

- An adequate system for the performance measurement of an information security framework.

- Continuous improvement of management system operations by discovering and correcting non-conformities as they arise.

An ISMS boosts an organization's likelihood of regularly achieving the important success criteria essential to safeguard its information assets.

The ISO 27000 series of standards cover all the requirements, including sector-specific ones for implementing a robust and sustainable ISMS. The organization chooses what to implement based on the business requirements.

The ISO 27000 series

Businesses of any kind can manage the security of assets such as financial information, intellectual property, employee details, or information entrusted by third parties by using the ISO/IEC 27000 family of standards. They cover a wide range of businesses, large and small, in every industry.

In response to changing information security requirements in many industries and contexts, new standards are being developed to keep pace with the rapid advancement of technology.

There are a number of standards in the ISMS family that do the following:

- Outline the standards for an ISMS and for those who certify such systems (for example, ISO/IEC 27001, ISO/IEC 27006, and so on)

- Assist in the whole process of establishing, implementing, maintaining, and improving an ISMS (for example, ISO/IEC 27002, ISO/IEC 27003, ISO/IEC 27004, and so on)

- Address industry-specific rules for the ISMS (for example, ISO/IEC 27010, ISO/IEC 27011, and so on)

- Deal with ISMS conformance assessment (for example, ISO/IEC 17021)

ISO 27001, and other management system standards published by ISO, undergo periodic reviews and updates to ensure their continued relevance and effectiveness in addressing emerging risks and evolving industry practices. These revisions reflect the commitment of the standard-setting bodies to incorporate advancements in technology, address emerging threats, and align with changing regulatory requirements to maintain the highest standards of information security management.

Let's look at a few of the ISO 27000 series of standards that have been published.

ISO/IEC 27001

This standard is known as **Information security, Cybersecurity, and Privacy protection – Information security management systems – Requirements** (https://www.iso.org/).

This standard talks about the requirements for implementing an effective information security management system. Using ISO/IEC 27001, an organization can build and operate an ISMS that includes a set of controls for controlling and mitigating risks connected with its information assets. Organizational conformance can be audited and certified.

One further set of criteria and guidelines for a **Privacy Information Management System (PIMS)** is specified in ISO/IEC 27701, which is an extension of ISO/IEC 27001 (ISMS).

All businesses of any kind and size may benefit from the standard since it helps them fulfill legal obligations while also managing privacy concerns associated with **Personally Identifiable Information (PII)**.

ISO/IEC 27006

This standard is known as **Information technology – Security techniques –Requirements for bodies providing audit and certification of information security management systems** (https://www.iso.org/home.html).

This standard lays out the requirements and offers guidance to organizations that do ISMS audits and certifications. Its primary purpose is to facilitate the accreditation of certifying organizations that issue ISMS certifications. Organizations that provide ISO/IEC 27001 audits and ISMS certification should follow this standard's criteria and recommendations.

ISO/IEC 27006 is a supplement to ISO/IEC 17021 that establishes the accreditation requirements for certification firms for them to provide compliance certifications that meet the ISO/IEC 27001 requirements.

ISO/IEC 27002

This standard is known as **Information security, Cybersecurity and Privacy protection – Information security controls** (https://www.iso.org/).

This standard establishes guidelines and management techniques for corporate information security. Using the standard's controls and best practice recommendations, implementers can make well-informed decisions about which controls to use and how to put them in place to fulfill their information security goals.

The ISO/IEC 27002 guideline is a code of practice for information security controls that outlines the procedures for implementing the security controls established in the ISO 27001 standard.

ISO/IEC 27003

This standard is known as **Information technology – Security techniques – Information security management systems – Guidance** (https://www.iso.org/).

ISO/IEC 27003 is intended to assist organizations in designing and implementing an ISMS. It gives straightforward instructions on how to plan an ISMS project in organizations of all sizes and sectors.

ISO 27001:2013 specifies the *what*, whereas ISO 27003 specifies the *how*. It provides direction for the actions required to implement and launch an ISMS.

ISO/IEC 27004

This standard is known as **Information technology – Security techniques – Information security management – Monitoring, measurement, analysis and evaluation** (https://www.iso.org/).

ISO/IEC 27004 specifies methods to evaluate ISO 27001's performance. The standard is designed to assist companies in assessing the efficacy and efficiency of their ISMS by providing the information essential for managing and improving the framework in a methodical manner.

Additionally, it defines how to develop and implement measurement processes, as well as how to evaluate and report on the results of connected measurement constructs, which enables the effectiveness of an ISMS to be evaluated in accordance with ISO/IEC 27001.

ISO/IEC 27005

This standard is known as **Information security, cybersecurity, and privacy protection – Guidance on managing information security risks** (https://www.iso.org/).

This standard contains risk management recommendations for information security and is intended to aid in the successful implementation of information security using a risk management strategy. An efficient ISMS must identify organizational needs in relation to information security requirements and follow the guidelines in ISO/IEC 27005, which explains how to carry out a risk assessment in compliance with ISO/IEC 27001 criteria.

For an organization, risk assessments are critical to the ISO/IEC 27001 compliance process.

ISO/IEC 27007

This standard is known as **Information security, cybersecurity, and privacy protection – Guidelines for information security management systems auditing** (https://www.iso.org/).

This standard gives advice on performing ISMS audits and on the competence of auditors. In order to administer an ISMS audit program in accordance with the requirements defined in ISO/IEC 27001, businesses must follow ISO/IEC 27007.

ISMS audit program management, auditing, and the competency of ISMS auditors are all addressed in these guidelines. They may be used by anybody who needs to understand or perform an ISMS audit, whether it's internal or external, or who needs to manage an ISMS auditing program.

ISO/IEC TS 27008

This standard is known as **Information technology – Security techniques – Guidelines for the assessment of information security controls** (https://www.iso.org/).

This standard contains instructions for conducting a review and assessment of information security controls. These controls are evaluated in accordance with an organization's established ISMS framework. This document offers guidance on how to review and assess how well the controls have been implemented, how they are working, and how well they have been technically evaluated.

Information security assessments and technical compliance checks are relevant to all kinds and sizes of organizations, including public and private businesses, government agencies, and not-for-profit ones.

ISO/IEC 27013

This standard is known as **Information security, cybersecurity, and privacy protection – Guidance on the integrated implementation of ISO/IEC 27001 and ISO/IEC 20000-1** (https://www.iso.org/).

This standard provides guidance to users on how to establish a dual management system that includes procedures and documentation. This will guide the deployment of ISO/IEC 27001 and ISO/IEC 20000-1 simultaneously or sequentially and match the current ISO/IEC 27001 and ISO/IEC 20000-1 management system specifications.

As a result, businesses are better able to design an integrated management system that complies with both ISO/IEC 27001 and ISO/IEC 20000-1 standards and comprehend the features, similarities, and differences between the two.

ISO/IEC 27014

This standard is known as **Information security, cybersecurity, and privacy protection – Governance of information security** (https://www.iso.org/).

An organization's information security actions can be evaluated, directed, monitored, and communicated using ISO/IEC 27014. According to the guidelines laid forth in this standard, information security governance should be based on principles and processes. Information security management may be assessed, directed, and monitored with the use of this. If an organization's information security measures are breached, it may have a negative effect on the organization's public image. A requirement of this standard is that an organization's governing bodies be given oversight of information security to guarantee that its objectives are fulfilled.

ISO/IEC TR 27016

This standard is known as **Information technology – Security techniques – Information security management – Organizational economics** (https://www.iso.org/).

This standard lays forth the principles by which an organization should make decisions regarding the security of its data by considering the financial impact of such decisions. The technical report equips the organization with the knowledge necessary to more accurately assess the risks associated with its identified information assets, comprehend the value that information security measures add to those assets, and determine the appropriate level of resources to apply to secure those assets.

It outlines how an organization may make information-protection choices and assess the economic implications of such decisions in the setting of conflicting resource demands.

ISO/IEC 27010

This standard is known as **Information technology – Security techniques – Information security management for inter-sector and inter-organizational communications** (https://www.iso.org/).

This standard establishes guidelines for information security collaboration and coordination between organizations within the same domain, between domains, and with authorities. When it comes to inter-organizational and inter-sector communications, this standard provides guidelines for the implementation of information security management. It also provides controls and guidance related to the inception, implementation, maintenance, and improvement of information security in those communications.

The guidelines apply to all types of sensitive information transmission and sharing (public and private, national and international, within the same sectors or across industry sectors).

ISO/IEC 27011

This standard is known as **Information technology – Security techniques – Code of practice for Information security controls based on ISO/IEC 27002 for telecommunications organizations** (https://www.iso.org/).

This standard offers information security management recommendations for telecommunications businesses. The ISO/IEC 27002 rules have been adapted to fit the needs of the industrial sector.

ISO/IEC TR 27015

This standard is known as **Information technology – Security techniques – Information security management guidelines for financial services** (https://www.iso.org/).

In addition to the recommendations provided in the ISO/IEC 27000 family of standards, ISO/IEC TR 27015 offers guidance for establishing, implementing, maintaining, and enhancing information security in financial services companies.

ISO 27017

This standard is known as **Information technology – Security techniques – Code of practice for information security controls based on ISO/IEC 27002 for cloud services** (https://www.iso.org/).

ISO 27017 is a collection of principles for securing cloud-based infrastructures and reducing the risk of security incidents. Customers may be confident that a business is committed to providing secure cloud services and that it has procedures in place to deal with any difficulties that may arise as a result of that commitment.

ISO 27018

This standard is known as **Information technology – Security techniques – Code of practice for protection of personally identifiable information (PII) in public clouds acting as PII processors** (https://www.iso.org/).

The ISO/IEC 27018 standard is a set of rules or a code of conduct for the selection of PII protection measures as part of the implementation of an ISO/IEC 27001-based cloud computing information security management system.

ISO 27799

This standard is known as **Health informatics – Information security management in health using ISO/IEC 27002** (https://www.iso.org/).

This standard includes guidelines for establishing an ISMS to help healthcare organizations in adopting an ISMS that has industry-specific adaptations of ISO/IEC 27002 standards.

Following the ISO 27000 series of standards helps organizations protect their critical and confidential data. In this section, we saw the various standards available in the ISO 27000 family. Although there are numerous standards in the family, only a few are relevant as such from an implementation perspective, which were explained.

Summary

This chapter discussed the family of information security standards that can be implemented to ensure the CIA triad across an organization. You also learned about the ISMS framework and its relevance and the ISO 27000 series of standards. In the next chapter, we will discuss the origin and structure of the ISO 27001 framework. You will also learn in detail about the PDCA cycle, legal and regulatory compliance, certifications and accreditations, and more.

2
Introduction to ISO 27001

The ISO framework is a set of policies and processes that organizations can use. ISO 27001 establishes a framework to assist organizations of any size or industry in protecting their information in a systematic and cost-effective manner by implementing an **Information Security Management System (ISMS)**. It is a differentiator for your business and shows other businesses that they can rely on your organization to manage valuable third-party information assets/data and intellectual property, which opens up a slew of new options while shielding your company from risks.

Companies can obtain ISO 27001 certification and demonstrate to their clients and partners that they protect their data, in addition to receiving the necessary know-how from the standard. Individuals can also be certified as a Lead Auditor or Lead Implementer of the ISO 27001 standard. It has global recognition as an international standard.

In this chapter, you'll learn about the process of accreditation and certification followed by the emergence of the ISO 27001 standard and its structure, as well as the trade-offs and benefits for a business entity in implementing an ISMS. We will also cover the legal side and regulations. The topics covered in this chapter are as follows:

- ISO 27001's origin
- ISO 27001's structure and PDCA
- Implementing ISO 27001 – a SWOT analysis
- Legal and regulatory compliance
- Accreditations and certifications

ISO 27001's origin

ISO 27001 is a global standard that originated with the establishment of management standard BS 7799. The standard was split into two parts:

- **Part 1**: A code of practice that deals with controls and provides formally managed information security. It was adopted as ISO 17799 *Information Technology – Code of Practice for Information Security Management* in December 2000.

- **Part 2**: A specification for implementing the ISMS. It was first published by the **British Standards Institution** (**BSI**) in 1999 as *Information Security Management Systems – Specifications with guidance for use*. Later, in 2002, it was revised to introduce the **Plan-Do-Check-Act** (**PDCA**) quality assurance model.

As a standard for guiding the creation and implementation of an ISMS, BS 7799 was developed with the support of the UK **Department of Trade and Industry** (**DTI**). The BSI published BS 7799 in 1995. A key goal of BS 7799 was to allow an organization's management to be confident in the effectiveness of its information security measures and arrangements even if the technology it used was proprietary. When BS 7799 was first developed, its primary goal was to safeguard corporate information in three ways: availability, confidentiality, and integrity. Importantly, however, it does not discuss protection against every possible threat, but only from those that the organization considers significant and only to the extent that a risk assessment is justified financially and economically.

Originally, BS 7799 was only a single standard, and it was regarded as a code of conduct. Instead of being created as a specification that could serve as the basis of an external third-party verification and certification method, it was more of a directive to companies. There was an increasing need for a certification option tied to the standard as more enterprises became aware of the scope, severity, and interconnection of information security risks, as well as the emergence of a growing body of data protection and privacy-related laws and regulations. Due to this, a second portion of the standard, designated BS 7799-2, was created as a specification (part 2).

ISO 17799 and BS 7799-1 have been assigned to the code of practice, which deals with controls rather than ISMSs (part 1). Additionally, a link was formed between the Code of Practice and the standard, with ISO 17799 mandating the use of ISO 17799 as the source of assistance in selecting and implementing the controls prescribed by ISO 27001. ISO adopted BS 7799-2 as ISO/IEC 27001 in November 2005 and ISO 27001 was established as the only certifiable standard for ISMS.

With the origin of this key standard discussed, let's move on to understanding its structure.

ISO 27001's structure and PDCA

The ISO 27001 standard defines the criteria for creating, implementing, and maintaining an organization's ISMS. The complete name of the standard is *Information security, cybersecurity and privacy protection — Information security management systems — Requirements* (as per the latest version released in 2022) and it consists of two parts:

- **The main part**: This consists of 11 clauses (0 to 10), in which clauses 0 to 3 describe the standard itself and clauses 4 to 10 describe the requirements your company must meet to be compliant with the standard

- **Annex A**: This consists of 93 controls that are to be considered while implementing ISMS

> **Annex L**
>
> For each management discipline, ISO has developed a management system standard. Although the technical content of each standard differs according to the relevant management discipline, ISO has developed a high-level framework structure (originally called Annex SL, and later, in 2019, renamed as Annex L), which provides the generic clause titles, text, common terms, and core definitions for a management system standard to be developed. Basically, Annex SL is a high-level structure for all future development of ISO standards to follow. It applies to every ISO standard, meaning that they will all have the same structure.
>
> As per the framework, the high-level structure follows this structure:
>
> 1. Scope
>
> 2. Normative references
>
> 3. Terms and definitions
>
> 4. Context of the organization
>
> 5. Leadership
>
> 6. Planning
>
> 7. Support
>
> 8. Operation
>
> 9. Performance evaluation
>
> 10. Improvement

According to W. Edwards Deming's PDCA cycle (see *Figure 2.1*), business processes should be regarded as though they are in a continuous feedback loop so that management may identify and adjust areas of the process that require improvement. Planning should be done first, followed by implementation and performance measurement. The measurements should be checked against the planned specification to identify any deviations or potential improvements, and they should be reported to management so that a decision can be made about the next course of action.

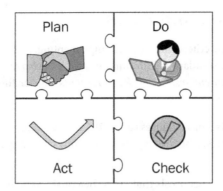

Figure 2.1 – PDCA cycle

A prerequisite of ISO 27001 (ISMS) is the PDCA process, which has its origins in quality assurance. PDCA needs to be performed before comprehending ISO 27001's requirement processes. PDCA analysis of ISO 27001 will provide you with a clearer picture of how governance implementation and alignment with improved business objectives will look.

Let's look, in detail, at the clauses of ISO 27001 and how they are aligned with PDCA here:

- **Clauses 0, 1, 2, and 3** are self-explanatory and act as metadata that gives general information about the standard:

- **Clause 0: Introduction** gives a general overview of the standard and its purpose and explains the compatibility with other ISO standards.

- **Clause 1: Scope** defines the scope of the standard and points out that this standard is applicable to all types of organizations.

- **Clause 2: Normative References** refers to content that is essential to understand and implement a certification standard (ISO 27001 in this case). These references provide additional guidance, requirements, or best practices that organizations should consider when implementing an Information Security Management System (ISMS) based on ISO 27001.

- **Clause 3: Terms and Definitions** refers to the terms and their definitions given in the ISO 27000 standard and is applicable to ISO 27001 as well.

Clause 2 and 3 refer to the ISO 27000 standard where terms and definitions are given (covered in the previous chapter).

Plan

The Plan phase consists of clauses 4 through 7. It helps to recognize an opportunity and plan a change:

- **Clause 4: Context of the organization** – Understanding the organization's context is a prerequisite for successfully implementing an ISMS (how and where it operates). This is done by analyzing the external and internal issues that influence the information security of your organization. Understanding the legal and regulatory requirements, the economic and political environment in which the company operates, and also the social and cultural norms contributes to knowing the external factors. To understand the internal issues, you need to know the organizational structure, culture, and values of the company. Next, determine the parties that are interested in information security in your company depending on the nature of the business. This can include clients, partners, suppliers, employees, and local authorities. Once the parties are identified, determine their requirements regarding information security. As diverse as the group, the requirements will vary accordingly. For example, some clients may give you sensitive or personal information, which might be bound by legal requirements, in which case they would require you to protect it appropriately. Next, from the obtained requirements from interested parties, plan and agree on the requirements that will be addressed through the ISMS that is to be implemented.

  Only with a thorough understanding of these factors can an effective ISMS be established in any organization.

- **Clause 5: Leadership** – Leadership commitment, information security policies, and organizational roles and responsibilities are the key components in this clause. Commitment by top management is mandatory for a management system. It is the foundation for establishing an information security policy and objectives. Other examples of obligations to meet include providing resources for the ISMS, including allocating people, time, and financial resources. This policy should be documented and communicated to all relevant parties, both within and outside of the organization. The ISMS should be integrated into the organizational processes and applied to the day-to-day activities. This sets a good example for the employees, and as a result, the ISMS will not be viewed as a separate entity but as a part of company operations. Implementation of the ISMS should be followed by steps for continual improvement. Management can enable employees of the organization to give feedback and propose improvements. Without proper management support, ISMS implementation in an organization would probably fail.

  The information security policy should include the intentions of the company regarding information security and should clearly show management's commitment to satisfying security requirements and continually improving the ISMS. It should also enable the establishment of information security objectives. The information security policy is the basis of the ISMS in any organization and gives it direction. As a top-level policy document, it will not include details of security controls.

Clause 5 also refers to the process of clearly defining and assigning specific roles and responsibilities within an organization. Defining and assigning roles and responsibilities for information security and communicating those to everyone in the organization lets the employees understand what is expected of them, what their impact is on information protection, and how they can contribute. The two types of responsibilities that top management assigns here are responsibilities that ensure that the ISMS is fully implemented and responsibilities for monitoring the performance of the ISMS and reporting to top management.

- **Clause 6: Planning** – Risks and opportunities should constantly be considered while making plans in an ISMS setting. Risks refer to unwanted events that can have a negative impact on the company. Opportunities refer to actions that the company can take in order to improve its information security. Identifying, documenting, and managing risks and opportunities are key to a successful ISMS because they help organizations see what the strengths and weaknesses of their business operations are and use them to build effective information security. The organization's risk assessment should be taken into account while establishing information security objectives. What an acceptable risk for the company is should be clear. Finally, a risk treatment plan is derived. Any update required for the ISMS is carried out in a planned manner.

- **Clause 7: Support** – A sufficient amount of competent resources, expertise, communication, and control over recorded information is crucial to supporting the cause. Without appropriate resources (including financial and human), it wouldn't be possible to run an ISMS. This is management's responsibility. The company should define the necessary skills to perform information security activities and ensure that employees have the required training and experience. External or in-house training and mentorships can help in upskilling the employees in this regard. Even if there are perfect documentation and controls in place, if people don't know how to put them into practice, then security will fail. It is also required that employees know what to do and why. Emails, newsletters, discussion groups, and online courses can help employees understand organizational security goals better.

Communicating information is the essence of understanding, and understanding what is happening with security is key to ISMS success. Here, what type of information is to be communicated inside and outside of the company and who is allowed/responsible for this should be determined. Rules for communication are framed based on the information security objectives of the organization.

Information must be documented, developed, updated, and controlled in accordance with ISO 27001 requirements. Information can range from providing guidance on how processes are conducted (policies, procedures, and so on) to evidence of activities conducted (records). It is essential that this information is protected and can be accessed when needed in a form suitable to use. There should also be clear identifiers, such as the reference number, title, date of creation, and author, for the information.

Do

Clause 8 is the Do phase. It aims to test the change:

Clause 8: Operation – The ISMS needs to be operated on a daily basis. The company implements numerous information security controls, processes, and actions for addressing risks and opportunities and makes sure everyone is complying with them. The implementation includes defining the criteria for processes and implementing the required controls as per the criteria. The information security policies and procedures need to be periodically reviewed so as to ensure changes in the company are reflected and accommodated in them. A change in the company can be intentional or unintentional. There should also be an owner for the process. Outsourced operations for an organization need to be identified and appropriately controlled with regard to information security.

Also, the information security risk assessment should be conducted regularly and at planned intervals. A first-time risk assessment activity can seem far more complex compared to the follow-up reviews. Once the assessment is done, the more strategic and costly task of risk treatment takes place. A thorough **Statement of Applicability (SoA)**, along with a comprehensive risk assessment and treatment methodology, will lay the groundwork for figuring out what to do about your security.

Check

Clause 9 is the Check phase. It reviews the test, analyzes the results, and identifies what you have learned.

Clause 9: Performance evaluation – The ISO 27001 standard expects the ISMS to be monitored, measured, analyzed, and evaluated. The case may be that the company decides "what needs to be measured" and is put in policies, objectives, and documented procedures. Mainly, the controls and security processes' performance are measured against the policies, objectives, and established procedures. The methods used for measurement and analysis must be defined to get suitable results. The results are to be presented to the top management. Management must review the ISMS of the company at planned intervals, considering the status of action items from past reviews, inputs from internal and external context, interested parties, risk assessment, and so on to ensure its effectiveness. Improvements such as the automation of processes can be adopted by management.

Act

Clause 10 is the Act phase. It is basically acting based on what you have learned in the Check phase.

Clause 10: Improvement – When a requirement is not complied with, it results in nonconformity. Nonconformities are to be addressed by taking corrective actions in a timely manner. ISO 27001 requires companies to keep records as evidence of the nature of the nonconformity, the actions that were taken, and the results of the implemented corrective actions. In addition to that, the company should also ensure that it has really resolved the root cause of the nonconformity. A procedure for corrective actions is to be documented and followed, as a good practice.

Moreover, continual improvement is an integral part of ISO 27001. The ISMS must be made more suitable, adequate, and effective.

An ISMS improvement policy can include the following:

- Defining who is responsible for planning, managing, and coordinating improvement activities
- Communicating that all employees can contribute to continual improvement
- Defining ways to record all relevant information related to improvement
- Implementing the improvement by documenting the changes and the rationale behind them and the expected outcome, and reviewing the effectiveness of the changes
- Improvement is a continuous process, wherein we constantly figure out what works the best.

The second part of the ISO 27001 standard, Annex A, comprises 93 controls grouped into four categories. An SoA defines which Annex A controls are applicable for the organization in context, by considering various factors such as the nature of the business and applicable laws and regulations. Defining the SoA is a crucial step in implementing the ISMS framework. It is a mandatory document and sits at the heart of your ISMS implementation.

However, the ISO 27001 standard, when implemented in an organization, has its downsides as well. A **Strengths, Weaknesses, Opportunities, and Threats** (**SWOT**) analysis will give a realistic, fact-based, data-driven look at the outcome.

Implementing an ISMS – a SWOT analysis

In strategic planning, the SWOT analysis framework is used to assess a company's competitive position and identify opportunities and threats (see *Figure 2.2*). A SWOT analysis takes into account both internal and external issues, as well as present and future opportunities.

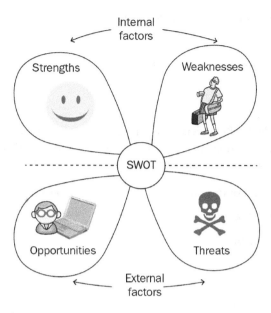

Figure 2.2 – SWOT analysis

Implementing an ISMS requires the organization to expend its resources and is a strategic decision. A SWOT analysis of implementing an ISMS can help in defining the organizational context and interested parties and ensure the effectiveness of the implementation.

It can be useful for developing a better understanding of the security environment and supporting the business by giving insights into security assets, risks, issues, and challenges that the IT department – and thus the business as a whole – will be faced with.

Let's check out the SWOT analysis for an organization through the implementation of an ISMS.

Strengths

The strengths of an ISMS implementation are as follows:

- Improves the overall security level of the organization
- Better portrayal of compliance with various regulations regarding data protection, thereby ensuring improved *return on investment*
- Differentiates the business in the eyes of the customer, in an increasingly competitive market
- Portrays the credibility of the business
- Defined life cycle to address weaknesses

- Better security for information (especially sensitive information)
- Brings good practices
- Lowers the expenses resulting from security incidents

Weaknesses

There are instances where the ISMS implementation can prove to be a weakness to the organization. The gap between the implementation and the desired result is to be analyzed for effective decision-making. Some of the major weaknesses can be as follows:

- Considerable investment of time and financial and human resources in the process
- Getting accustomed to the processes
- The handling of cost and the complexity of the ISMS if not properly designed and implemented

Opportunities

ISMS implementation can open new opportunities for the business, such as the following:

- Provide an edge in marketing
- Opportunities for improvement can be identified in the auditing process
- Better market positioning, thereby attracting better opportunities

Threats

Implementing an ISMS can pose threats related to security policy and practices, such as the following:

- Risk of revealing organizational information to third parties
- Can be resource-intensive
- Overconfidence in the ISMS protecting the entire information

In a SWOT analysis, a company, initiative, or sector's strengths and weaknesses are examined objectively using facts and statistics. There should be no preconceived notions or gray zones in the analysis, and instead, the focus should be on real-life situations. However, it should be used as a recommendation rather than a prescription by companies.

Although the benefits are greater than the drawbacks, before implementing the ISO 27001 standard, conducting a SWOT analysis gives more clarity to management on the aspects they need to focus on. Once implemented, the ISMS framework brings credibility along with greater responsibility for the organization in terms of following the legal and regulatory aspects.

Legal and regulatory compliance

In a dynamic world, laws and regulations are subject to change depending on a range of factors, such as the industry, country, and type of information. In some industries, companies are well versed in the rules, while in others, they are just learning about them for the first time. Because of the frequency and significance of security-related incidents, governments around the world have realized the importance of safeguarding people and businesses from the misuse of sensitive information. This is a constantly evolving topic since new laws and regulations are required as technology evolves.

Without adequate security measures, a company is vulnerable to cyberattacks, hefty fines or penalties imposed by authorities, legal action taken because of carelessness, and unwelcome media attention that could undermine the company's reputation, brand, and overall worth. There should be a clear definition and documentation of all statutory, regulatory, and contractual requirements as well as any system security needs. Legal and regulatory compliance in this regard is not the responsibility of the IT department or information security team alone. It is critical to have assistance from several key functions, including legal and contracts, human resources, and finance.

Depending on how you do business, it's not uncommon for certain organizations to be subject to several rules, including laws from multiple nations, some of which may have conflicting requirements. Creating an outline of all legislation and contractual responsibilities is best done in conjunction with your legal department (or even a professional expert). To assess whether your present security measures are sufficient for compliance or further steps are needed to satisfy the criteria, identify which requirements may have an impact on your company and then talk with your security staff about the results.

Consultation on applicable laws and regulations may also be obtained from a professional service. Despite the fact that regulatory compliance has a significant impact on information security, empowering the legal team to get familiar with laws and regulations, or employing professionals to do it for you, can save a significant amount of time and effort. Compliance efforts will lead to better security procedures and greater commercial value in the long run.

> **Conformance to ISO standards and legal requirements**
>
> Infractions of the law may result in a fine or other legal consequences. Failure to comply with an ISO standard when contractual requirements apply may result in a contractual violation, which may qualify for legal action.

The certification body verifies the compliance with regulatory requirements of an organization by verifying whether all the necessary controls are implemented and proper actions are taken in the case of non-compliance.

Accreditations and certifications

Obtaining ISO 27001 certification proves a company's dedication to continuous information asset/ sensitive data enhancement, development, and protection by putting in place suitable risk assessments, acceptable policies, and appropriate controls.

If an organization receives ISO 27001 certification, it indicates the organization's ISO 27001 ISMS has been audited and found to be in accordance with the standard by another entity known as a **certification body**. Organizations with ISO 27001 certification are announcing to the public that they are trustworthy, have implemented an ISMS in accordance with the standard, and have demonstrated compliance with the ISO certification body. Obtaining certification demonstrates to partners, stakeholders, and customers that your company takes information security management seriously.

The **International Accreditation Forum** (**IAF**) is the core authority that oversees the entire conformity assessment accreditation and related bodies that perform conformity assessment in the fields of management systems, products, services, and so on. It is a global organization comprising certification agencies from various countries. These standard-setting organizations are tasked with the responsibility of developing, modifying, and designing essential global standards. Membership to the IAF is open to accreditation bodies that conduct and administer validation/verification body accreditation programs and/or accreditation bodies for the certification of management systems and other conformity assessment programs, products, processes, services, employees, and so on (see *Figure 2.3*).

Examples are as follows:

- **United Kingdom Accreditation Service (UKAS)**: Accreditation body established in the UK
- **International Accreditation Service (IAS)**: Accreditation body established in the US
- **ANSI National Accreditation Board (ANAB)**: Accreditation body established in the US
- **National Accreditation Board for Certification Bodies (NABCB)**: Accreditation body established in India

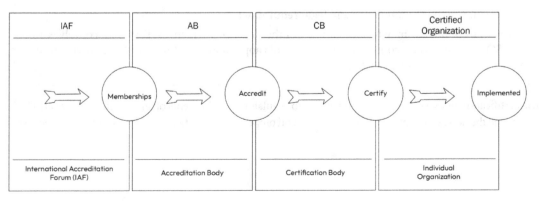

Figure 2.3 – Accreditation hierarchy

IAF is an association of accreditation bodies. The accreditation bodies select and accredit certification bodies, which in turn issue a certificate to an organization after conducting the ISMS audit.

Examples of certification bodies are BSI, TUV SUD, and Intertek.

The certification audit generally consists of two stages – Stage 1 and Stage 2.

- **Stage 1** is known as **Documentation Review**, where the auditor reviews your processes and policies to verify whether they are in line with the ISO standard.

- **Stage 2** is called the **Certification Audit**, in which the auditor assesses the organization's compliance with the ISO 27001 standard. Here, basically, the auditor evaluates the implementation effectiveness of ISO controls based on the defined SoA.

More details on the audit process will be discussed in *Part 3* of this book.

Summary

ISO 27001 evolved from the BS 7799 standard and garnered global recognition. ISO/IEC 27001 itself is divided into two parts, the first one comprising 11 clauses and the second one (Annex A) 93 controls. The clauses are categorized into phases of Plan, Do, Check, and Act for implementation. The decision to implement an ISMS by an organization can be planned better with the SWOT analysis tool. After implementation, the responsibility of adhering to legal and regulatory compliance is a priority. The accreditation and certification processes are the hierarchical steps in achieving the certificate. This chapter will have helped you understand the importance of analyzing the present status and context of the organization to implement an ISMS.

In the following chapter, we'll go into the specifics of each clause, including the Annex A controls.

Part 2:
The Protection Strategy – ISO/
IEC 27001/02 Design
and Implementation

Part 2, consisting of *Chapters 3* to *7*, delves into the heart of the ISO/IEC 27001/27002 standards, offering an in-depth understanding of controls, risk management, and the developmental stages of an ISMS. *Chapter 3* navigates through the controls laid out in ISO 27001/27002, illustrating how they can be interpreted and applied based on business context. *Chapter 4* dives into risk assessment and management and the crucial elements of the ISO 27001 framework, and introduces the role of a risk register. In *Chapter 5*, you'll journey through the process of developing an ISMS, learning how to customize control implementation to a business's specific context. *Chapter 6* underscores the significance of a comprehensive incident management plan to maintain information security. Finally, *Chapter 7* provides practical insights via real-world case studies concerning certification, a **Statement of Applicability (SoA)**, and incident management. Together, these chapters arm you with the skills to interpret, implement, and manage the ISO/IEC 27001/27002 standards in your organization.

This part has the following chapters:

- *Chapter 3, ISMS Controls*
- *Chapter 4, Risk Management*
- *Chapter 5, ISMS – Phases of Implementation*
- *Chapter 6, Information Security Incident Management*
- *Chapter 7, Case Studies – Certification, SoA, and Incident Management*

3
ISMS Controls

The ISO 27001 standard recommends taking a risk-based approach to information security. Organizations must identify and address information security threats by establishing controls as a result of this.

The measures are detailed in Annex A of the standard. In Annex A of the ISO 27001 standard, there are 93 controls separated into 4 groups – A.5 through A.8. The implementation of all 93 controls is not required, and only a small number of them are mandatory to be recorded. It is up to the company to determine what to implement and what not to, based on their risk management methodology. This freedom of choice allows businesses to focus on the controls that are most important to them rather than wasting money on those that aren't. The applicable controls are defined in the **Statement of Applicability (SoA)**.

ISO 27001's Annex A simply provides a one-sentence description of each control, giving you a sense of what the goal is or what needs to be accomplished, but not how to execute it. ISO 27002 provides a detailed description.

The following is a high-level breakdown of what each category of control focuses on:

- A.5 – organizational controls
- A.6 – people controls
- A.7 – physical controls
- A.8 – technological controls

ISO 27001 versus 27002

ISO/IEC 27002 provides guidance on how to implement the security controls in Annex A of ISO 27001, the international standard for an **Information Security Management System (ISMS)**. The ISO 27000 series is a collection of documents pertaining to various aspects of information security management. ISO 27001 is the fundamental framework in which the implementation requirements for an ISMS can be found. Basically, this is a list of everything you need to do in order to be compliant. Although ISO 27002 is a more comprehensive standard, no organization can be accredited to it as it is not a management standard. It is a collection of requirements for businesses to manage their policies and processes so that they can achieve a certain set of outcomes. It essentially lays down the rules for operating a system. In the case of ISO 27001, it defines the ISMS, and therefore certification against ISO 27001 is possible.

Information security must be designed, executed, monitored, reviewed, and enhanced as part of this management system. It entails that the top management of the organization has a specific set of tasks, such as setting and measuring goals and conducting internal audits. ISO 27001 specifies all of these aspects, but not ISO 27002.

Finally, ISO 27002 does not distinguish between controls that are appropriate to a specific company and those that are not, but ISO 27001 does. A risk assessment is required to determine whether or not each control should be implemented, as well as to what extent, in accordance with ISO 27001's guidelines.

So, why are these two standards kept apart rather than combined? Because only then are they usable. If it were a single standard, it would be far too complex and big to be useful.

ISO 27001 is for laying the foundation and creating a framework for information security in your organization; ISO 27002 is for implementing controls; ISO 27005 is for conducting risk assessments and risk management; and so on. Each standard in the ISO 27000 series serves a specific purpose in the information security field.

ISO 27001 cannot be implemented without ISO 27002, and ISO 27002 would remain an isolated effort by a few information security enthusiasts without the management framework provided by ISO 27001, and thus would have no real impact on the organization's security.

Annex A controls

The Annex A controls are from A.5 through A.8. There are 4 control sets/domains, and a total of 93 controls in Annex A. Let's see in detail what each control domain in Annex A is.

A.5 – organizational controls

Organizational controls include the management and governance structure that supports and guides the implementation and operation of information security within the organization. This structure encompasses the roles, responsibilities, policies, and procedures that create the organizational backbone for achieving the company's information security goals. The main aim of these controls is to establish standards, measure performance, compare performance to those standards, and take necessary corrective action to ensure that the organization's security posture aligns with its objectives and policies. This category emphasizes the importance of leadership, strategic planning, and organizational structure in ensuring comprehensive information security.

The controls in this domain are listed as follows:

Control Reference	Control Name	Control Description
5	Organizational Controls	
5.1	Policies for information security	Establishes the requirement for defining, communicating, and regularly reviewing management-approved information security policies that are acknowledged by relevant personnel and interested parties
5.2	Information security roles and responsibilities	Ensures the establishment and distribution of information security-related roles and responsibilities as per the specific requirements of the organization
5.3	Segregation of duties	Emphasizes the need for separating roles and responsibilities within an organization to prevent potential conflicts and protect the integrity of operations
5.4	Management responsibilities	Mandates that all personnel adhere to and implement information security in line with the organization's established policies and procedures
5.5	Contact with authorities	Mandates that the organization establishes and maintains communication with relevant authorities to ensure compliance and facilitate cooperation in legal and security matters
5.6	Contact with special interest groups	Mandates the creation and maintenance of relationships with specialist security forums and professional associations for ongoing learning and collaboration on information security matters
5.7	Threat intelligence	Involves the systematic collection and analysis of information about security threats to generate actionable intelligence that can guide protective measures

Control Reference	Control Name	Control Description
5.8	Information security in project management	Ensures the integration of information security measures and principles into all stages of project management
5.9	Inventory of information and other associated assets	Mandates the creation and maintenance of a detailed inventory of all information and associated assets, along with their respective ownership information
5.10	Acceptable use of information and other associated assets	Establishes guidelines for the acceptable use of the organization's information and associated assets to ensure they are used appropriately and securely
5.11	Return of assets	Specifies procedures for the secure return of organizational assets upon termination or change of employment, contractual agreement, or any other situation where assets need to be returned
5.12	Classification of information	To categorize information based on the organization's security needs considering aspects of confidentiality, integrity, availability, and requirements of relevant stakeholders
5.13	Labeling of information	To implement procedures for appropriately labeling information, based on the organization's adopted information classification scheme
5.14	Information transfer	To establish rules, procedures, or agreements for securely transferring information within the organization and between the organization and external parties
5.15	Access control	To establish rules for both physical and logical access to information and related assets, based on business and information security needs
5.16	Identity management	To govern the full life cycle of user identities, from creation to deactivation, to ensure appropriate access and security

Control Reference	Control Name	Control Description
5.17	Authentication information	To define a management process for the allocation and handling of authentication information, offering guidance to personnel on appropriate practices
5.18	Access rights	The process of providing, reviewing, adjusting, and revoking access permissions to information and associated assets according to the organization's specific access control policies and rules
5.19	Information security in supplier relationships	To establish processes and procedures to manage the information security risks associated with the use of suppliers' products or services
5.20	Addressing information security within supplier agreements	To establish and agree on relevant information security requirements with each supplier, based on the nature of the supplier relationship
5.21	Managing information security in the **Information and Communication Technology (ICT)** supply chain	To establish processes and procedures to manage the information security risks associated with the supply chain of ICT products and services
5.22	Monitoring, reviewing, and change management of supplier services	To enable regular monitoring, reviewing, evaluation, and management of changes in supplier information security practices and service delivery to maintain robust organizational security
5.23	Information security for use of cloud services	To establish procedures for the secure acquisition, usage, management, and discontinuation of cloud services in line with the organization's information security requirements
5.24	Information security incident management planning and preparation	To plan and prepare for managing information security incidents by defining and communicating the relevant processes, roles, and responsibilities

Control Reference	Control Name	Control Description
5.25	Assessment and decision on information security events	Evaluate information security events to determine whether they should be classified as security incidents
5.26	Response to information security incidents	Establishes the requirement for responding to information security incidents according to predefined and documented procedures
5.27	Learning from information security incidents	Implement a process to learn from and leverage knowledge gained from past security incidents to improve and strengthen the organization's information security controls
5.28	Collection of evidence	Establish procedures for the identification, collection, acquisition, and preservation of evidence related to information security events
5.29	Information security during disruption	Have strategies and procedures in place to maintain appropriate levels of information security during unforeseen disruptions or emergencies
5.30	ICT readiness for business continuity	Need to plan, implement, maintain, and test ICT readiness based on business continuity objectives and requirements
5.31	Legal, statutory, regulatory, and contractual requirements	Identify, document, and maintain up-to-date information about all legal, statutory, regulatory, and contractual requirements relevant to information security in the organization
5.32	Intellectual property rights	Procedures are necessary to protect the intellectual property rights of the organization and others
5.33	Protection of records	Measures are to be implemented to safeguard records from loss, destruction, falsification, and unauthorized access or release
5.34	Privacy and protection of **Personally Identifiable Information (PII)**	Establish guidelines for preserving privacy and protecting PII in accordance with applicable laws, regulations, and contractual requirements

Control Reference	Control Name	Control Description
5.35	Independent review of information security	Regular independent reviews of the organization's information security management approach and its implementation, including people, processes, and technologies, particularly when significant changes occur
5.36	Compliance with policies, rules, and standards for information security	Mandates regular reviews to ensure adherence to the organization's information security policy, specific topic policies, rules, and standards
5.37	Documented operating procedures	Information processing facilities' operations need to be documented and made accessible to relevant personnel

Table 3.1 – A.5 – organizational controls

The incorporation of organizational controls in the ISO 27002:2022 version is a significant step in highlighting the importance of management and governance in information security. It emphasizes the need for a strong organizational backbone, comprising well-defined roles, responsibilities, policies, and procedures that align with the company's information security objectives. These controls allow for the establishment of standards, performance measurement, and comparison, and the implementation of corrective actions when necessary. By fostering a culture of security through strategic planning and leadership, organizational controls ultimately help to create a resilient security posture that can proactively manage and mitigate information security risks. This category demonstrates that a robust information security framework goes beyond technology and requires a comprehensive and coordinated organizational effort.

In the next section, we will look, in detail, at the people controls.

A.6 – people controls

Given that human error or malicious intent can often be the weakest link in the security chain, people controls are designed to embed security-conscious behavior and practices at all levels of the organization. From background checks on new hires and outlining security responsibilities in employment contracts to providing ongoing security training and setting up procedures for reporting security incidents, people controls work to establish a robust culture of information security. In addition, these controls address contemporary challenges, such as remote working, emphasizing their relevance and adaptability to evolving work environments. By embedding these measures, organizations can better protect their information assets, thus ensuring business continuity, maintaining customer trust, and complying with regulatory requirements.

Each control under the people controls domain is listed in this table, with a description:

Control Reference	Control Name	Control Description
6	People Controls	
6.1	Screening	The control mandates conducting pre-employment and ongoing background checks on all prospective and existing personnel, in accordance with legal, regulatory, and ethical considerations, tailored to the organization's business needs, the sensitivity of accessible information, and the potential risks involved
6.2	Terms and conditions of employment	Information security responsibilities of both the organization and its personnel should be specified within employment contracts
6.3	Information security awareness, education, and training	Mandates the provision of appropriate information security awareness, education, and training, along with regular updates on the organization's information security policies and procedures relevant to individual job functions
6.4	Disciplinary process	A formally established and communicated procedure should be in place for taking action against individuals or parties that violate the organization's information security policies
6.5	Responsibilities after termination or change of employment	Define, enforce, and communicate ongoing information security responsibilities and duties that continue after employment termination or role change
6.6	Confidentiality or non-disclosure agreements	The establishment, documentation, regular review, and signing of confidentiality or non-disclosure agreements by staff and relevant parties are required to ensure the organization's information protection needs are met
6.7	Remote working	Security measures must be implemented to safeguard information accessed, processed, or stored by personnel working remotely, outside the organization's premises

Control Reference	Control Name	Control Description
6.8	Information security event reporting	Establish a system for personnel to promptly report observed or suspected information security incidents via the appropriate channels

Table 3.2 – A.6 – people controls

The people controls in ISO 27001 emphasize the vital role of human elements in information security management.

These controls cover eight key areas:

- Conducting background checks on new personnel (screening)

- Defining personnel and the organization's responsibilities for information security in employment contracts (terms and conditions of employment)

- Providing regular information security awareness, education, and training (information security awareness, education, and training)

- Implementing a formal disciplinary process for policy violations (disciplinary process)

- Detailing responsibilities and duties post-termination or employment change (responsibilities after termination or change of employment)

- Executing confidentiality or non-disclosure agreements (confidentiality or non-disclosure agreements)

- Instituting security measures for remote working (remote working)

- Establishing a system for reporting security events (information security event reporting)

These measures aim to enhance the integrity, confidentiality, and availability of information by managing and mitigating human-associated risks.

The next section discusses the physical controls.

A.7 – physical controls

Physical controls are the tangible and environmental measures taken to protect an organization's assets and resources, including data. These controls range from setting up barriers such as walls, doors, or locks to creating controlled zones with access restrictions. They also involve designing spaces to maintain the safety of individuals and the physical systems that store, process, and transmit information. The goal of physical controls is to prevent unauthorized access, damage, theft, and interference with organizational assets. They are a critical part of a comprehensive security program, as they provide a first line of defense against potential threats that could compromise the integrity, availability, and confidentiality of information. Implementing physical controls, as outlined in the ISO 27001 standard, can significantly improve an organization's overall security posture.

Each control that comes under the physical controls domain and its description are detailed in the following table:

Control Reference	Control Name	Control Description
7	Physical Controls	
7.1	Physical security perimeters	Define boundaries that protect areas containing important information and related assets
7.2	Physical entry	Protect secure areas through appropriate entry controls and access points
7.3	Securing offices, rooms, and facilities	Design and implement physical security for these spaces
7.4	Physical security monitoring	Continuous monitoring of premises should be done to detect unauthorized physical access
7.5	Protecting against physical and environmental threats	Design and implement protection measures against threats such as natural disasters and other physical threats to infrastructure
7.6	Working in secure areas	Implement security measures for operations in secure areas
7.7	Clear desk and clear screen	Enforce rules for clean desks and screens to maintain information security
7.8	Equipment siting and protection	Secure placement and protection of equipment
7.9	Security of assets off-premises	Ensure protection of assets located off-site
7.10	Storage media	Handling storage media throughout its life cycle, from acquisition and use to transport and disposal, based on the organization's classification scheme and handling requirements
7.11	Supporting utilities	Protection of information processing facilities from power failures and other disruptions caused by failures in supporting utilities

Control Reference	Control Name	Control Description
7.12	Cabling security	Protection of cables carrying power and data from interception, interference, or damage
7.13	Equipment maintenance	Proper maintenance of equipment must be done to ensure the availability, integrity, and confidentiality of information
7.14	Secure disposal or reuse of equipment	Verify that sensitive data and licensed software have been removed or securely overwritten before the disposal or reuse of equipment

Table 3.3 – A.7 – physical controls

Physical controls not only define measures for restricting physical access to secure areas but also dictate the protocols for secure equipment handling, protection against environmental threats, and the maintenance of secure working conditions. By regulating areas from utilities support to equipment disposal, these physical controls form a comprehensive shield against potential physical breaches, thereby upholding the integrity, confidentiality, and availability of the organization's information assets. Furthermore, they bolster the resilience of the organization, enabling it to respond and recover effectively from unforeseen disruptions or physical incidents.

A.8 – technological controls

Technological controls provide a structured and comprehensive framework for managing and securing an organization's digital infrastructure. These controls encompass various aspects of information technology, including user endpoint devices, data access restrictions, capacity management, malware protection, and secure software development. The importance of technological controls cannot be overstated in today's digital age, where information is increasingly processed, stored, and communicated electronically. The measures covered under these controls not only aim to safeguard against cyber threats but also ensure the optimal utilization of resources, secure data management, and the maintenance of system integrity. By implementing these controls, organizations can mitigate risks associated with technical vulnerabilities, prevent data breaches, and maintain the confidentiality, integrity, and availability of their information assets.

Each technological control and its description are listed in the following table:

Control Reference	Control Name	Control Description
8	Technological Controls	
8.1	User endpoint devices	Protect information that is stored on, processed by, or accessible via end user devices
8.2	Privileged access rights	Manage and restrict the allocation and use of privileged access rights
8.3	Information access restriction	Restrict access to information and other associated assets according to the established access control policy
8.4	Access to source code	Manage read and write access to source code, development tools, and software libraries
8.5	Secure authentication	Implement secure authentication technologies and procedures based on the policy of information access restrictions
8.6	Capacity management	Monitor and adjust the use of resources in line with current and expected capacity requirements
8.7	Protection against malware	Implement measures to protect against malware, supported by appropriate user awareness
8.8	Management of technical vulnerabilities	Obtain information about technical vulnerabilities and evaluate the organization's exposure to such vulnerabilities to take appropriate measures
8.9	Configuration management	Establish, document, implement, monitor, and review configurations, including security configurations, of hardware, software, services, and networks
8.10	Information deletion	Ensure information stored in information systems, devices, or other storage media is deleted when no longer required
8.11	Data masking	Use data masking in accordance with the organization's access control policy, other related policies, and business requirements

Control Reference	Control Name	Control Description
8.12	Data leakage prevention	Apply data leakage prevention measures to systems, networks, and other devices that process, store, or transmit sensitive information
8.13	Information backup	Maintain and regularly test backup copies of information, software, and systems
8.14	Redundancy of information processing facilities	Ensure that information processing facilities have sufficient redundancy to meet availability requirements
8.15	Logging	Produce, store, protect, and analyze logs that record activities, exceptions, faults, and other relevant events
8.16	Monitoring activities	Monitor networks, systems, and applications for anomalous behavior to evaluate potential information security incidents
8.17	Clock synchronization	Ensure synchronization of the clocks of information processing systems with approved time sources
8.18	Use of privileged utility programs	Restrict and tightly control the use of utility programs that can override system and application controls
8.19	Installation of software on operational systems	Procedures and measures should be in place to securely manage software installation on operational systems
8.20	Network security	Secure, manage, and control networks and network devices to protect the information in systems and applications
8.21	Security of network services	Identify, implement, and monitor security mechanisms, service levels, and service requirements of network services
8.22	Segregation of networks	Segregate groups of information services, users, and information systems in the organization's networks

Control Reference	Control Name	Control Description
8.23	Web filtering	Manage access to external websites to reduce exposure to malicious content
8.24	Use of cryptography	Define and implement rules for the effective use of cryptography, including cryptographic key management
8.25	Secure development life cycle	Establish and apply rules for the secure development of software and systems
8.26	Application security requirements	Identify, specify, and approve information security requirements when developing or acquiring applications
8.27	Secure system architecture and engineering principles	Establish, document, maintain, and apply principles for engineering secure systems
8.28	Secure coding	Apply secure coding principles to software development
8.29	Security testing in development and acceptance	Define and implement security testing processes in the development life cycle
8.30	Outsourced development	Direct, monitor, and review activities related to outsourced system development
8.31	Separation of development, test, and production environments	Separate and secure development, testing, and production environments
8.32	Change management	Changes to information processing facilities and information systems should be carried out in accordance with change management procedures
8.33	Test information	Select, protect, and manage appropriate test information
8.34	Protection of information systems during audit testing	Ensure that audit tests and other assurance activities involving the assessment of operational systems are planned and agreed upon between the tester and appropriate management

Table 3.4 – A.6 – technological controls

Technological controls aim to ensure the secure operation of an organization's digital systems, protect information from unauthorized access, and maintain system integrity. The wide-ranging and comprehensive nature of these controls reflects the complex and multifaceted nature of information security in our digital era. They extend beyond the simple protection of data, incorporating elements of access management, resource optimization, vulnerability assessment, and more. Implementing these controls can significantly help organizations navigate the modern digital landscape safely and effectively, providing robust defense mechanisms against cyber threats and ensuring the continuity of their operations. The consistent monitoring, reviewing, and updating of these technological controls are paramount to adapting to the evolving landscape of information security threats.

Summary

ISO 27002 is a code of practice for the controls of an ISMS, and it goes into far more detail than ISO 27001's Annex A controls. The list of 93 controls, divided into 4 control sets, is expanded in clauses A.5 to A.8 of ISO 27002:2022.

While ISO 27002 is not a certifiable standard in and of itself, following its information security management principles will help the company satisfy ISO 27001 certification requirements. It explains how to comply with the ISO 27001 standard and how to implement it.

Because there is no one-size-fits-all information security solution, the appropriate information security controls must be decided on based on their risk assessment and appropriate controls. The CIA triad can be used to define information security in this context.

In the upcoming chapter, we will see one of the most significant aspects of the entire information security auditing process – risk management. The important steps of identifying, analyzing, measuring, treating, and monitoring a risk are all explained there.

4

Risk Management

Managing risks while using information technology is known as **information security risk management**. It is the process of recognizing, assessing, and addressing threats to an organization's assets' confidentiality, integrity, and availability. Several coordinated actions can be used to lead and regulate a company's risk management.

The eventual goal of this approach is to treat risks that are beyond the organizational risk appetite as per the overall risk tolerance of a business. Rather than aiming for zero risk, companies should aim for a level of risk that is manageable for their company.

Managing risk is one of the most challenging aspects of implementing *ISO 27001*, but risk assessment (and treatment) is also the most critical phase at the beginning of any information security project since it lays the groundwork for your company's information security program.

When it comes to the ISO 27001 standard, it's all about determining which incidents are most likely to occur (assessing the risk) and then devising the best ways to avoid them (treating the risks). ISO 31000 and ISO 27005 are two standards that address risk management aspects in an organization. ISO 31000 is a globally recognized standard for risk management. While it is not a mandatory requirement for ISO 27001 implementation, ISO 31000 provides valuable guidance and best practices that can significantly enhance the effectiveness of an organization's **information security management system (ISMS)**.

The ISO 27005 standard focuses on information security risk management, while ISO 31000 provides guidelines for managing risks in general. ISO 27005 offers guidance for information security risk assessment and treatment within the context of ISO 31000's risk management framework.

This chapter covers the following main topics:

- ISO 31000:2018
- ISO 27005:2022

First, we will look into the main standard – that is, ISO 31000 – in detail.

ISO 31000:2018

To achieve ISO 27001 compliance, a company needs to have a solid risk management framework. Compliance can be achieved by implementing *ISO 31000* or a comparable risk management system. It's crucial for making sure that the ISMS that comes out of adopting the standard deals with the risks thoroughly and suitably.

The importance of ISO 31000 lies in its comprehensive approach to risk management. It provides a systematic and structured framework for identifying, assessing, treating, and monitoring risks across an organization. By adopting ISO 31000, organizations can establish a proactive risk management culture and make informed decisions to protect their information assets.

In 2009, the **International Organization for Standardization** (ISO) issued *ISO 31000*, an international standard that serves as a guide for the design, implementation, and maintenance of risk management; it was revised in 2018. The standard is named *ISO 31000:2018 Risk Management – Guidelines* (`https://www.iso.org/`). A wide range of factors and influences can make it difficult for any organization to achieve its goals, regardless of its size or scope. An organization's objectives are affected by this uncertainty, which is known as risk.

An organization's activities inherently carry a degree of risk. Risk management is described in *ISO 31000:2018* in terms of a systematic and logical process in which organizations identify and analyze the risk to determine whether risk treatment is necessary to satisfy their risk criteria.

An organization's various functions, projects, and activities can all benefit from risk management.

Concepts and generic guidelines are provided by ISO 31000 to assist enterprises in creating and operating their risk management system.

It can be utilized by any public, private, or community-based organization, association, group, or individual. All aspects of a company's operations, procedures, projects, goods, services, and assets can be measured according to this standard at any time during the organization's life.

Figure 4.1 shows the relationships between the risk management principles, framework, and process:

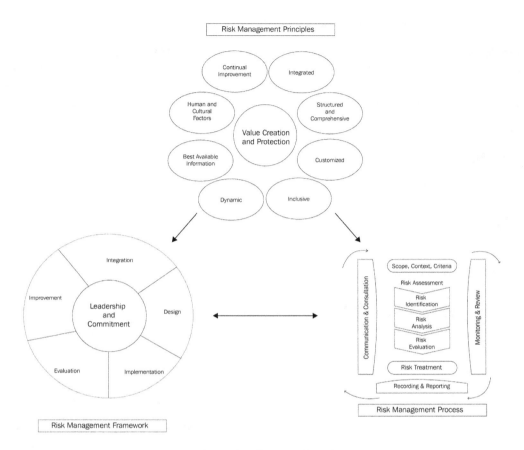

Figure 4.1 – Principles, framework, and process

Next, we'll look at the three clauses in ISO 31000.

Key clauses of ISO 31000:2018

ISO 31000 is organized into the following main clauses:

- **Clause 4**: Principles
- **Clause 5**: Framework
- **Clause 6**: Process

The upcoming sections describe each of them in detail. First, we will cover the risk management principles that are explained in clause 4.

Risk management principles

Principles of risk management are described in clause 4 and are intended to provide guidelines on the characteristics of efficient and effective risk management. Risk management frameworks and processes should be built around these principles, which are the foundations of risk management. These principles should make it possible for a company to deal with the impact of uncertainty regarding its goals.

These risk management principles are outlined in *Figure 4.2*:

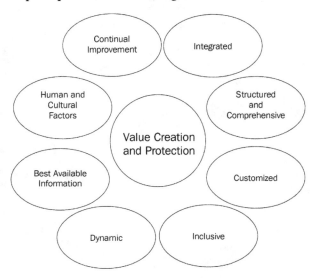

Figure 4.2 – Risk management principles

ISO 31000:2018, Risk Management – Guidelines, provides the following eight principles for a solid risk management program (see *31000-2018, Section 4, Principles*):

1. Integration
2. Structured and comprehensive
3. Customized
4. Inclusive
5. Dynamic
6. Uses best available information
7. Considers human and cultural factors
8. Practices continual improvement

Let's look at what each one means:

- **Integration**

  The risk management efforts of an organization should be integrated into all aspects of the organization.

- **Structured and comprehensive**

  When it comes to risk management, the best results are achieved when a methodical, systematic, and timely approach is developed and implemented.

- **Customized**

  An organization's approach to risk management must take into account not only its objectives but also the external and internal environments in which it works.

- **Inclusive**

  Every stakeholder must be involved in the process of risk management for it to be most effective. As a result, risk management initiatives can take into account the knowledge, perspectives, and perceptions of all stakeholders.

- **Dynamic**

  Organizations should adapt their risk management programs and initiatives to reflect changes in their external and internal environments. Successful businesses have learned to deal with the fact that change is inevitable. A sound risk management program can help organizations better anticipate, recognize, and respond to change.

- **Best available information**

  To perform effective risk management, it is necessary to take information from the past, the present, and the future into account. When it comes to assessing risk, risk managers need to take into account both the limitations and uncertainties. All key parties must receive timely and clear information.

- **Human and cultural factors**

  You must understand the human and cultural context in which the risk management effort is taking place, as well as the impact that these elements will have on the risk management endeavor. For example, a company transitioning to a new software system risks operational disruption and data loss due to employee resistance or errors, exacerbated by factors such as resistance to change, unfamiliarity with new technologies, and the organization's culture around change management. To mitigate these risks, the company can communicate the need and benefits of the change, provide thorough training, involve employees in the transition, and offer ongoing support. This approach considers and addresses the human and cultural factors that influence risk.

- **Continual improvement**

  Risk managers must constantly improve their organizations' risk management efforts through experience and learning.

These risk management principles guide the risk management framework. Now, let's look at the risk management framework in detail, which is the template that's used by organizations to identify, eliminate, and minimize risks.

Risk management framework

Clause 5 of the risk management framework outlines the company's strategy for incorporating risk management into its most critical operations and services. The effectiveness of a company-wide risk management framework is critical to a successful risk management strategy's implementation at all levels.

The risk management framework does the following:

- Helps the risk management process be used successfully

- Ensures that information concerning risk obtained from the risk management process is reported properly

- Validates that this data is used as a basis for decision-making and accountability at all relevant organizational levels

Figure 4.3 shows the different components of a risk management framework:

Figure 4.3 – Risk management framework

Leadership and commitment

To demonstrate a long-term commitment to risk management, the organization's management team must define a risk management policy and objectives, ensure legal and regulatory compliance, make sure resources are given to risk management, and allocate responsibilities at all levels.

An organization's risk management must be in line with the context of the organization, giving effective values to the organizational objectives and strategies and creating opportunities for top management to achieve the organization's objectives. Having a solid risk management framework in place helps the organization align its risk management strategies with its objectives, ensure that the company and its stakeholders are aware of the established risk criteria, communicate the risk management strategy within the organization and to relevant stakeholders, and ensure that risks are tracked methodically.

Integration

Depending on the organization's structure and context, risk is integrated and managed at each level of the organizational structure.

The rules, policies, and practices necessary to realize the organization's goals are governed by the top management. To attain the appropriate levels of sustainable performance and long-term viability, management structures translate governance directives into a strategy and its associated objectives. An organization's governance includes determining who is responsible for risk management and who is responsible for overseeing risk management. A company's management will be able to focus its resources on risks that affect the attainment of the organization's objectives, safeguard assets, assure continuity of the organization's activities, and take effective decisions with ease if it implements integrated risk management.

This process is continuous and iterative, and it should be tailored to the needs and culture of the company in question. The organization's purpose, governance, leadership and commitment, strategy, objectives, and operations should all be integrated with risk management.

Design

To design a risk management framework, the following steps must be adhered to:

- **Understand the organization and its context**: The external context must include a wide range of technological, socioeconomic, and environmental factors, which can be international, national, regional, or local. The internal context can include the organization's vision, mission, strategies, contracts, interconnections, interdependencies, and so on.

- **Establish a risk management policy**: Having a risk management policy in place demonstrates the commitment of top management toward risk management. This commitment is demonstrated in ways such as integrating risk management at all levels of the organization, making the necessary resources available, measuring and reporting an organization's performance indicators, continual improvement, and so on.

- **Establish responsibility, authority, and competence for risk management**: The roles and responsibilities concerning risk management must be assigned to individuals (risk owners) and communicated as a core responsibility.

- **Provide the necessary resources**: It is the responsibility of top management to allocate appropriate resources for risk management. This includes human resources, tools and procedures, information and knowledge management systems, meeting training and upskilling needs, and so on.

- **Establish internal and external communication and consultation**: There should be timely communication and consultation and organizations should ensure that information reaches the intended audience. The stakeholders must also be provided with channels of feedback so that they can contribute to decision-making.

Once the framework has been designed, the organization can implement it.

Implementation

The framework for risk management and the risk management process must be implemented by the organization. This takes into account the timeline and resources, decision-making processes, checkpoints, and so on. The risk management strategy of the organization should be clearly understood and practiced. Any uncertainty in the process should be acknowledged and addressed. When appropriately designed and implemented, the framework ensures that risk management is part of all the activities throughout the organization and its decision-making.

Evaluation

Risk management must be evaluated periodically to see whether it is effective, as well as to check whether the policies and plans in place to implement it are still relevant. This evaluation helps ensure that the risk management framework is constantly aligned with the objectives of the organization.

Continual improvement

Based on the results of the evaluation, gaps and improvement opportunities are identified, based on which action plans to improve the risk management framework are implemented so that it stays effective. The organization can adapt the risk management framework to accommodate internal and external changes and address relevant gaps to improve the framework.

The next section explains the risk management process, which includes activities such as **risk assessment**, **risk treatment**, devising a risk treatment plan, and more.

Risk management process

Risk management is the critical first step in implementing *ISO 27001*. It determines everything that happens during the implementation of controls. Risk management is the process of dealing with risks, including activities such as identifying relevant risks, assessing and treating them, and so on.

From an asset-based to a process-based approach

An asset-based risk management approach was followed by organizations for a long time. However, at the time of writing, ISO recommends a process-based approach. The asset-based approach includes identifying the assets of the company and the related threats and vulnerabilities of those assets. Hardware, software, information (in various forms, such as electronic and paper), infrastructure, people, and so on are all assets of an organization. For example, a fire or earthquake can harm hardware assets. Similarly, malware can impact software systems negatively. Later, ISO began recommending a process-based approach with the introduction of the ISO 9001:2000 standard for quality management systems. This means looking at the ISMS not just as a set of individual elements but as a series of interrelated processes that work together to achieve the overall goals of the system. It is an essential approach recommended by ISO to perform effective audits and to ensure a complete and functioning ISMS.

The process-based approach was a key innovation of this version of the standard, replacing the previous emphasis on a more procedure-based approach. The goal was to provide a more holistic and flexible framework, based on the understanding that an organization's output is the result of a system of interrelated processes. Since then, the process-based approach has been incorporated into many other ISO standards, including ISO 27001 for information security management systems. The process-based approach is not just for auditing and risk management, but also for designing and implementing the ISMS itself. This allows organizations to design their ISMS around their unique business processes and risks, rather than having to fit their operations into a predefined structure.

Let's look at the main activities in risk management.

Communication and consultation

At each level of the risk management process, it is essential to communicate and consult with all relevant parties, both internal and external. It is important to convey the risks involved and the actions to be taken to all stakeholders. Communication increases awareness, while consultation allows you to receive feedback regarding risk management; this should take place throughout the risk management process. The result will be better accommodation of diverse expertise and views in defining risk criteria and risk evaluation, a sense of inclusiveness and ownership among those affected by risk, and better decision-making.

Establishing the scope, context, and criteria

Defining the scope of the risk management process involves stating the objectives, expected outcomes, specific inclusions and exclusions, tools and techniques used, resources required, roles and responsibilities, dependencies with other projects, and so on.

Understanding the internal and external context of the organization is important because the risk management process is interrelated with the objectives and activities of the organization. The organization should establish the external and internal context of the risk management process by considering the sociocultural, regional, legal, and regulatory frameworks, stakeholders, contractual commitments, vision and mission, organizational culture, and so on.

Figure 4.4 shows the structure of the risk management process:

Figure 4.4 – Risk management process

To have successful risk management, companies must define the steps and rules for managing risks. There are mainly five steps in the risk management process:

1. Risk assessment methodology
2. Risk assessment
3. Risk treatment
4. Statement of applicability
5. Risk treatment plan

Let's look at each step in detail.

Risk assessment methodology

Define a risk assessment methodology that will ensure consistent results when the assessment is done throughout the company. The methodology must be documented and must set the criteria for conducting the information security risk assessment and criteria for accepting risks. In reality, risks cannot be reduced or avoided completely and hence the company must define its risk appetite, which means what represents an acceptable risk for the company. For example, the absence of access control in the cafeteria of a company might be an acceptable risk, while a data center at the same company having no access restrictions is not an acceptable risk.

Risk assessment

In the context of information risk management, a risk assessment helps organizations assess and manage incidents that have the potential to cause harm to sensitive data.

The process involves identifying vulnerabilities that an attacker could exploit or mistakes that employees could make. It is the process of combining the information you've gathered about assets, vulnerabilities, and controls to define risk. The purpose of risk assessment is to help the company identify information security risks, as well as to analyze and evaluate the level of risks. The first step is to figure out potential problems that could occur at your company. You should take a look at all of your company's assets, as well as the threats and vulnerabilities they face, and then calculate the level of risk associated with each combination of these factors.

Once the level of risk has been determined, you can decide on the best course of action to avoid it.

Risk treatment

After assessing the risks, the next step is to select appropriate risk treatment options. For each unacceptable risk, the company must identify a response strategy.

There are mainly four strategies to choose from when it comes to risk treatment:

- **Risk reduction/modification**: This is one of the most common strategies. It can be achieved by implementing the relevant security controls from *Annex A* of *ISO 27001*.
- **Risk sharing**: This involves sharing or transferring the risk with a third party – for example, buying an insurance policy.
- **Risk avoidance**: This involves not pursuing the activities that cause the risk. In such cases, the impact of the risk is too high to be managed by other options. If the risk of unauthorized access to laptops in your organization is quite high, it is better to avoid taking them out of company premises.
- **Risk acceptance**: The organization accepts the risk and decides not to take action. This is only preferable if the cost of treating it is greater than its impact if realized.

The decision of which strategy to choose and implement is up to management.

Statement of applicability

The **statement of applicability (SoA)** is a mandatory document that includes a comprehensive list of all the security controls in place. It identifies the controls that an organization has chosen to address the risks, explains why these were selected, whether the organization has implemented these controls, and why any controls were excluded. The SoA is a window into the organization's controls.

First and foremost, an auditor will check the scope of an ISMS, which is a criterion for the ISO certification of an ISMS. Irrespective of the size of the company, the SoA and scope will cover all of your products and services, as well as your information assets, processing facilities, systems in use, and the people involved in your business.

In a well-presented and understandable SoA, Annex A controls and the risks and information assets under consideration are clearly stated. It demonstrates that the company is serious about information security management.

Risk treatment plan

Creating a **risk treatment plan (RTP)** is the last step in the process and is a mandatory document that's required by the standard. Once the SoA is in place (in which the applicable controls are specified), a risk treatment plan can be created. The RTP contains the controls, the people responsible for them, the financial and human resources required, and the timeline. It should be comprehensive and include the following:

- Planned actions, proposed priorities, or timetables
- Resource requirements
- The roles and responsibilities of all parties participating in the proposed actions
- Performance metrics
- Reporting and monitoring needs

All stakeholders' values and views should be reflected in action plans (for example, internal organizational units, outsourcing partners, customers, and so on). The more effectively the suggested plans are conveyed to the various stakeholders, the easier it will be to get approval and commitment to their implementation.

RTP is an action plan that helps the company implement all the selected controls to treat risks.

The next section will explain the risk assessment and mitigation (or treatment) strategies. Assessment is about risk identification and analysis, whereas mitigation is about reducing the impact of risk.

Assessment and mitigation

The steps to eliminate a risk includes its assessment and mitigation, in serial order. According to *ISO 27001*, a risk assessment should include elements such as risk identification, risk analysis, and, finally, a risk evaluation (shown in *Figure 4.5*):

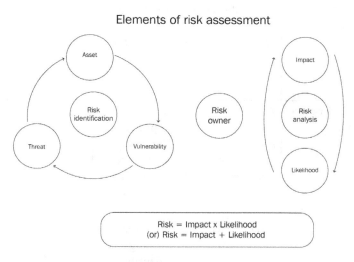

Figure 4.5 – Elements of a risk assessment

Risk assessment is the first step, followed by mitigation. Let's see what the risk assessment process is and the steps involved.

Risk assessment

The process of risk assessment mainly involves **risk identification** and **risk analysis**.

The best approach to risk assessment is based on the ISO 27005 standard for information security risk management. Asset-based risk assessment is a commonly used method here. It should be noted that although this asset-based approach has been traditionally followed, recent revisions of the ISO standards recommend a process-based approach. To prepare a risk assessment table, threats, vulnerabilities, and values of impact and likelihood are identified for each risk.

Identifying threats involves recognizing what could happen to assets that could have a negative impact. Identifying vulnerabilities means introspecting the reason those threats could happen. For example, a lack of antivirus software can make way for a virus attack, or a lack of an incident response procedure can cause problems if a bomb threat occurs.

Risk identification is an important step in the risk management process because to evaluate or treat the risks, first, you need to know which risks exist. ISO 27001 requires the risk assessment results to be documented.

The following is a sample risk assessment table based on the process-based approach:

Risk	Risk Owner	Threat	Vulnerability	Impact (1-5)	Likelihood (1-5)	Risk Value (= I+L)
Unauthorized access to sensitive data due to weak password policies	IT security manager	Cyber attackers using brute-force or guessing attacks	Weak password policy enforcement			
Data loss due to hardware failure	IT operations manager	Aging hardware, natural disasters, or unexpected equipment failure	Lack of equipment redundancy and absence of a robust data backup system			
Phishing attacks leading to compromised systems	IT security manager/ HR (for staff training)	Cyber attackers using phishing techniques	Employees not being able to identify and report phishing emails			

Table 4.1 – Risk assessment table

The impact and likelihood values are filled based on the analysis, and the risk value is calculated accordingly:

- **Risk owner**

    For each identified risk, a risk owner must be assigned. The risk owner is the person responsible for dealing with a particular risk, including activities such as approving any risk treatment plans, accepting the level of residual risk, and similar. Individual risks should be owned by the members of an organization who end up using their budget to pay for fixing the problem. In other words, risk owners are accountable for ensuring risks are treated accordingly.

- **Residual risk**

    The risk that remains after treatment is referred to as residual risk. After identifying the risks, those that are above the acceptable level are mitigated (that is, treated). Because no risk can be totally eliminated, some will still exist. This is known as the residual risk.

- **Risk analysis**

  Analyzing the identified risks includes assessing the impact of the risk on the asset and the likelihood of the identified risk happening. The impact here is examined based on the confidentiality, integrity, and availability of information. For example, if a server is down and the information in it cannot be accessed, the availability is impacted. Similarly, if there is unauthorized access to stored information, confidentiality is compromised. There are different scales to choose from to determine impact and likelihood, and it is up to the organization to make a choice. The scale can be high-medium-low, 0 – 10/0 – 5, and so on. The decisions must be documented in the risk assessment methodology.

  The level of risks can be determined using the following formula:

  Risk = Impact x Likelihood or Risk = Impact + Likelihood

  The formula that's used must also be documented in the risk assessment methodology.

- **Risk evaluation**

  First, the organization needs to arrive at a decision on which risks to accept and which not, based on the established methodology. The risks that are not acceptable need to be treated in the risk management step.

Based on the decision-making regarding the acceptability of risks, the risk assessment table is populated with values for calculated risks:

Risk	Risk Owner	Threat	Vulnerability	Impact (1-5)	Likelihood (1-5)	Risk Value (= I+L)
Unauthorized access to sensitive data due to weak password policies	IT security manager	Cyber attackers using brute-force or guessing attacks	Weak password policy enforcement	4	5	9
Data loss due to hardware failure	IT operations manager	Aging hardware, natural disasters, or unexpected equipment failure	Lack of equipment redundancy and absence of a robust data backup system	3	3	6
Phishing attacks leading to compromised systems	IT security manager/ HR (for staff training)	Cyber attackers using phishing techniques	Employees not being able to identify and report phishing emails	4	5	9

Table 4.2 – Risk assessment table with calculated risk values

After the assessment, the next step is risk mitigation. The following section explains what mitigation is and the sequential steps in the process.

Risk mitigation

Risk mitigation is the practice of anticipating disasters and devising a strategy for mitigating their negative consequences. While risk mitigation is based on the principle of preparing an organization for all possible hazards, an effective risk mitigation plan would consider the impact of each risk and prioritize planning around that impact. Rather than attempting to avert a risk, mitigation focuses on the aftermath of a disaster and the efforts that can be taken before the occurrence of the event to mitigate its detrimental and potentially long-term repercussions. In an ideal world, an organization would be prepared for and immune to all risks and hazards. However, having a risk mitigation strategy in place can help an organization prepare for the worst-case scenario by accepting that some level of damage is inevitable and putting processes in place to address it. When developing a risk mitigation strategy, a few procedures are generally conventional for most organizations. Recognizing recurrent risks, prioritizing risk mitigation, and monitoring the implemented plan are critical components of a comprehensive risk mitigation strategy.

The process of risk mitigation consists of the following steps:

1. **Prioritize risks**, which entails assigning a severity rating to measured risks. Prioritization is one facet of risk mitigation; it involves accepting some risk in one area of the organization to better safeguard another. By setting an acceptable degree of risk for certain areas, an organization can better plan for **business continuity** (**BC**) while putting fewer mission-critical business processes on hold.

2. **Track risks** involves monitoring risks as their severity or relevance to the organization changes. It is critical to have robust metrics for monitoring risk as it changes and for monitoring the mitigation plan's ability to comply with regulatory obligations (for example, **risk register**).

3. **Risk treatment method**, which is selected depending on the type and nature of the risk. There are four ways that organizations can treat risks:

   - **Risk reduction/modify the risk**: This is one of the most common strategies for risk treatment and is achieved by applying security controls to reduce the likelihood of it occurring and/ or damage it will cause.

   - **Avoid the risk**: You can avoid risk by changing the circumstances that are causing it – that is, simply stop performing the tasks or processes that incur such risks that are simply too big to mitigate with any other options. An example of this is banning the usage of laptops outside of the company premises if the risk of unauthorized access to those laptops is too high (because such hacks could result in the catastrophic failure of the complete IT infrastructure).

- **Share the risk**: This involves sharing the risk with a partner, such as an insurance company or a third party, that is better suited to manage the risk. To put it another way, if you acquire an insurance policy against fire for your home, you're transferring some of your financial risk to the insurance provider. This alternative would have no impact on the underlying risk.

- **Retain the risk/accept the risk**: In this case, your organization accepts the risk without doing anything about it. This is the least ideal choice. Only if the mitigation cost is greater than the damage an incident would cause should this option be employed.

The controls in ISO 27001 Annex A (and any other controls that a company thinks are appropriate) are used to reduce risks, which is the most typical approach to risk management.

4. **Implement and monitor progress**, which involves evaluating the mitigation plan's effectiveness at recognizing risk and making necessary adjustments. After implementing a strategy, it should be tested and analyzed regularly to ensure that it remains current and functional. Because the risks to data centers are always changing, risk mitigation plans must adapt to any changes in risk or shifting priorities.

The identified risks can be addressed in a variety of ways to reduce their impact. These possibilities are evaluated, and then action plans are formulated and put in place. Priority should be given to mitigating the most serious threats.

Choosing the best risk management strategy entails weighing the costs and rewards of each activity. In general, risk management expenses should be proportional to the advantages they provide.

In the next section, we will learn about ISO 27005, which is a more specialized extension or application of ISO 31000, and provide guidance and best practices specifically for information security risk management within the broader risk management context outlined by ISO 31000.

ISO 27005:2022

ISO 27005 was released in June 2008; the latest revision was released in 2022. The standard is named *ISO 27005:2022 Information Security, Cybersecurity, and Privacy Protection — Guidance on Managing Information Security Risks* (https://www.iso.org/). It assists organizations in identifying, assessing, and treating risks related to information security.

Although the exact steps may vary based on organizational context, ISO 27005:2022 outlines the following major steps, all of which can be followed in the information security risk management process:

1. **Establish the context**: The first step in the ISO 27005:2022 risk management process is to establish the context. This involves defining the scope of the risk assessment, identifying relevant stakeholders, and understanding the organization's objectives, constraints, and risk tolerance levels. By establishing the context, organizations can lay the foundation for effective risk management.

2. **Identify information security risks**: The next step is to identify information security risks. This involves systematically identifying potential threats, vulnerabilities, and impacts on the organization's information assets. Organizations can use various techniques, such as risk scenario concepts and asset-based approaches, to identify and document information security risks.

3. **Analyze information security risks**: Once the risks have been identified, they need to be analyzed. This step involves assessing the likelihood and potential consequences of each identified risk. Organizations can use quantitative or qualitative methods to analyze the risks and prioritize them based on their significance and potential impact on the organization.

4. **Evaluate information security risks**: After analyzing the risks, the next step is to evaluate them. Risk evaluation involves comparing the analyzed risks against predetermined criteria, such as risk appetite or tolerance levels. This step helps organizations determine whether the risks are acceptable or require further treatment.

5. **Treat information security risks**: Once the risks have been evaluated, organizations need to develop and implement risk treatment plans. Risk treatment involves selecting and applying appropriate controls and measures to mitigate or manage the identified risks. Organizations should consider a range of options, such as implementing security controls, transferring risks through insurance, or accepting risks based on informed decisions.

6. **Monitor and review**: The final step in ISO 27005:2022's risk management process is to monitor and review the effectiveness of the implemented risk treatment measures. This involves regularly assessing the performance of controls, monitoring changes in the risk landscape, and reviewing the risk management process itself. Organizations should continuously improve their risk management practices based on lessons learned and changing circumstances.

It is important to note that ISO 27005 aligns with other standards such as ISO/IEC 27001 and ISO 31000, which provide a comprehensive framework for information security management and general risk management, respectively.

Alignment with ISO/IEC 27001

ISO/IEC 27005:2022 has been revised so that it aligns better with ISO/IEC 27001:2022, which is the standard for ISMSs. The guidance text and terminology in ISO/IEC 27005:2022 have been adjusted to align with the layout and structure of ISO/IEC 27001:2022. This alignment ensures that organizations can effectively integrate information security risk management practices into their overall information security management systems.

In conclusion, ISO 27005 serves as a vital international standard for information technology risk management. It helps organizations rationalize sensitive data protection, anticipate the consequences of cyberattacks and cybercrimes, and integrate risk management processes with ISO/IEC 27001 and ISO/IEC 27002. By following the guidelines of ISO 27005, organizations can effectively manage information security risks and protect their valuable assets.

Summary

Risk management is about decision-making and taking actions to address uncertain outcomes, controlling how risks might impact the achievement of business goals.

ISRM is the process of identifying, evaluating, and treating risks around the organization's valuable information. **Confidentiality, integrity, and availability** (**CIA**) are the criteria against which the information assets are assessed for risks. Risk management aims to address uncertainties around those assets to ensure the desired business outcomes are achieved.

ISO 31000:2018 is a recently updated version of the **International Organization for Standardization** (**ISO**) standard for risk management that defines risk as *the effect of uncertainty on objectives.*

Managing risk is an ongoing task, and its success will come down to how well risks are assessed, plans are communicated, and roles are upheld. Identifying the critical people, processes, and technology to help address these steps will create a solid foundation for a risk management strategy and program in your organization, which can be developed further over time.

5
ISMS – Phases of Implementation

An **information security management system** (**ISMS**) comprises the various policies, standards, procedures, practices, behaviors, and scheduled activities that a corporation implements to protect the (important) information assets it possesses. Both the organization and its external constituents are provided with clear objectives and context regarding information security.

The design and implementation of the ISMS are dependent on the organization's requirements and goals. The organization's size and structure, the market or service region, and the sensitivity of the information it possesses or controls on behalf of others should also be considered. It is the goal of an ISMS to identify, assess (if necessary), and manage information security threats, to protect an organization's digital assets. This procedure shouldn't be considered a one-time event but an ongoing risk management cycle. The measurement and reporting of control efficacy are facilitated by the ISMS. Successful implementations of the ISMS offer a framework that is both realistic and pragmatic for the identification and management of information security threats.

Management, employees, suppliers, and regulatory authorities come under the scope of the ISMS and represent critical resources. A company's implementation of an ISMS is a strategic business choice. Keeping this in mind, it is critical to acknowledge that top management's buy-in and cooperation are necessary for a successful ISMS implementation.

It is important that when components of the ISMS are produced, they are deployed and used, rather than aiming for a *big-bang* style implementation. Also, the pursuit of perfection need not be the goal. The ISMS has mechanisms for both self-correction and improvement.

There is no requirement that an ISMS is based on the **ISO 27001** standard; however, this standard does provide a framework that is globally recognized and comprehended. This same structure also makes it possible to obtain an ISMS certification, which is acknowledged all around the world. An ISMS that is built on the ISO 27001 standard takes a comprehensive, hierarchical, and integrated approach to the process of detecting and addressing information security risks. It entails taking into consideration concerns pertaining to policy and procedure, the technology and tools that are utilized, and most crucially, people and the behavior of individuals.

Let us explore the phases of implementation of an ISMS. There are 16 different phases of implementation that we will explore in the following sections.

We will cover the following main topics in the chapter:

- Phases of ISMS implementation
- Time, effort, and roles in an ISO 27001 implementation

Phases of ISMS implementation

The following sections cover a step-by-step explanation of the various aspects, in sequential order, of ISMS implementation, based on the ISO 27001 standard.

1) Management support

Convincing management about an ISMS implementation can seem a daunting task. After all, management's ultimate responsibility is the profitability of the company and decisions will be based on **ROI** (short for **return on investment**). Planning how to present the information in a way that management can understand and endorse is one of the key aspects of convincing them.

It is obvious that management will look for the benefits of the proposed ISMS. The following are the four most important benefits of an ISMS:

- **Compliance**: ISO 27001 can provide a methodology that enables a company to comply with multiple regulations concerning data protection, privacy, and IT governance (particularly if it is an organization in the financial sector, the healthcare industry, or the government) in the most efficient and cost-effective way possible.

- **Marketing edge**: When competing in a crowded market, having ISO 27001 on your resume may help you stand out from the crowd, especially if potential customers demand that their data be handled carefully.

- **Reducing expenses**: In most cases, information security is viewed as an expense with no clear indication that it will result in any financial return. However, there is potential for monetary gain if you reduce the expenses brought about by security incidents. There is a good chance that you have service outages, as well as the occasional loss of data. There is not yet a mechanism or a piece of technology that can quantify how much money you could save if you were able to prevent instances of this kind. However, bringing such incidents to the attention of management almost always results in a positive response.

- **Bringing order to your business**: If your company has grown significantly over the past several years, you may have encountered problems such as trying to discern who is in charge of information assets and who is allowed access to information systems. Implementing ISO 27001 will drive you to define roles and responsibilities in a very detailed way, which will enhance your internal structure and be particularly helpful for sorting these things out.

In conclusion, ISO 27001 may offer more than just a certificate to hang on your wall. In most circumstances, management will start paying attention to you if you explain these benefits in a concise manner.

2) Accomplishing a project

Implementing an ISMS that is based on ISO 27001 is a complex task that involves many people and a wide range of activities. It can take anywhere from a few months (for smaller companies) to more than a year (for large corporations).

The key to successful implementation is to figure out what needs to be done, who will do it, and when (that is, apply project management).

3) Defining the scope

The primary objective of determining the scope of the ISMS is to define the types of information that are going to be protected. Therefore, it makes no difference where this information is stored – on your company's premises or in the cloud – or how it is accessed – from your local network or from a distance. The most important thing to keep in mind is that you are still in charge of protecting this information regardless of how, where, or by whom it is accessed.

Therefore, just because your employees use computers outside of the office does not mean that you are not responsible for them. Instead, if employees can access your local network and all the sensitive data and services that are housed there from their own devices, these devices should be covered under your scope of work.

The scope considers the interested parties, logical boundaries, physical boundaries, and exclusions. It specifies the bounds within which the management system operates. In most cases, the scope encompasses the entirety of the organization, including all its processes, products, and services. To define the scope, it is necessary to have knowledge of the budget, the legal requirements, the requirements of the standard, the intention of top management, the current state of the organization, and so on. It is possible to carry out a gap analysis to determine what components of the management system are lacking in the organization. This can be useful in determining whether there is a disconnect between what the organization does and what else is necessary for a management system to be compliant. Additionally, it enables the management of available resources.

The scope will be utilized for the purpose of identifying the targets of the risk assessments, thereby providing support for the remaining elements of the ISMS. It might also not be required to implement an identical level of security throughout the entire organization. A poorly defined scope is one of the factors that can contribute to the failure of an ISMS implementation. It is necessary to have a formal, documented definition of the scope of the ISMS in place, regardless of what the scope is.

When drafting an ISMS scope, the following need to be addressed:

- The mission of the company
- The goals of the ISMS
- The business activities of the organization
- What is vital and should be safeguarded, for example, intellectual property and customer information
- The physical scope of the project
- Is the ISMS applicable to all business units?
- The inclusion of service providers
- The presence of exclusions

When the organization has little to no management authority over that region, it may result in exclusions, making it challenging to manage any risks.

As per ISO 27001, when defining the scope, the following must be considered:

- The internal and external issues (the context) of the organization *[1]*

> **Internal and external context**
> Internal context refers to variables that are within the company's control (e.g., the structure of the organization); external context refers to circumstances beyond an organization's control that it must prepare for and adjust to (e.g., political and economic conditions).

- The needs and expectations of interested parties (relevant stakeholders interested in your ISMS) *[2]*

> **Interested parties**
> Employees – due to the fact that they adhere to the ISMS's prescribed procedures – and customers – since they make use of your services and trust you with private data.

- The connections and interdependencies that exist between the ISMS and the outside world *[3]*

> **Dependencies and interfaces**
>
> Dependencies are the services offered by parties outside of your scope (e.g., legal services). Interfaces help define the ISMS's boundaries and show what data will be flowing in and out.

A smaller company may choose to include the entire organization in the scope as it is easier, while a larger company may implement ISO 27001 for only a part of the organization. ISO 27001 requires the ISMS scope to be documented – it can be merged with some other document (e.g., an information security policy) or kept as a separate document.

An example of the scope statement is as follows: *"The design, development, and maintenance of firmware services offered to customers with the support of the R&D division located in Texas."*

4) Information security policy

The **information security policy** is the highest-level document in your ISMS, and it should not be overly detailed but should outline some basic information security requirements for your organization. Policy documents are intended to help management make clear its goals and how it intends to achieve them. Policies such as the **user acceptance policy**, **physical security policy**, and **email policy** will be created to complement the enterprise's information security strategy.

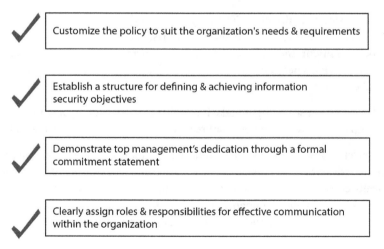

Customize the policy to suit the organization's needs & requirements

Establish a structure for defining & achieving information security objectives

Demonstrate top management's dedication through a formal commitment statement

Clearly assign roles & responsibilities for effective communication within the organization

Figure 5.1 – Factors to consider in framing an ISO 27001 information security policy

There are many factors to consider when creating an information security policy, as shown in *Figure 5.1* and listed here:

- The policy needs to be adapted to the organization – the policy of a large manufacturing company may not fit a small IT company.

- An information security policy must define how objectives are proposed, approved, and reviewed to establish the framework for it.

- The policy must demonstrate top management's commitment to meeting the needs of all stakeholders and continually improving the ISMS.

- It is excellent practice to designate a person who will communicate the policy internally and externally (for example, customers and suppliers), and that person should do so on a regular basis.

- Regular (for example, annual) reviews of the policy are required; the policy's owner should be identified and held accountable for keeping the policy up to date.

Since ISO 27001 mandates that management guarantees that the ISMS and its objectives are aligned with the company's long-term strategic goals, the information security policy should serve as a primary link between your top management and your information security activities (clause 5.2 of ISO 27001). To be effective, the policy needs to be kept short and understandable to top management.

5) Risk assessment methodology

When it comes to assessing risk, the ISO 27001 project's most difficult work is defining the rules for risk identification and determining an acceptable level of risk. Decide whether you want a qualitative or quantitative **risk assessment**, the scales you'll use for the former, the level of risk that's acceptable, and other factors. The risk owner should always make the judgment regarding the risk level (its consequence and likelihood).

A quantitative risk assessment scale might make it possible to look at an organization's risk profile in more detail and set up a standard to see how mature it is when the right security controls are in place.

The project manager/chief information security officer can handle risk assessment and treatment on their own or with the assistance of an outside specialist (for example, a consultant). The risk assessment procedure is typically carried out by the same project team that is responsible for implementing ISO 27001 at larger companies. Smaller companies may want to complete the process themselves, with the support of proper documentation and tools, and not opt for external help.

6) Risk assessment and risk treatment

Risk assessment can take several days for a small business, and it might take several months for larger ones, so you need to plan your efforts meticulously in this step. The goal is to gain a complete view of the threats to your company's data from both internal and external sources.

ISO 27001 requires five main steps for risk assessment:

1. **Risk identification**: This is basically listing assets, threats, and vulnerabilities. Digital tools can be used to list assets, threats, and vulnerabilities in columns; information such as the risk ID, risk owners, impact, and likelihood also need to be included.

2. **Assigning risk owners**: The person(s) responsible for risk is identified based on their knowledge of the asset and/or their power to make necessary changes to the asset.

3. **Risk analysis**: Assessing consequences (impact) and their likelihood.

4. **Risk calculation**: This is the determination of the level of risk, done by either addition or multiplication of impact and likelihood. For example, say the impact is 4 and the likelihood is 2 (on a scale of 1 to 5, where 1 = least value and 5 = highest value). The level of risk can be *8 (4x2)* or *6 (4+2)*. The method to be adopted (addition or multiplication) is decided by the organization's risk management framework. Note that one method should be adopted uniformly across all risk calculations

5. **Risk evaluation**: Accepting the risks based on criteria set/defined in the organization's risk assessment methodology. If the value obtained in risk calculation is *8*, but the acceptable value is *5*, that means the risk is not acceptable.

To reduce unacceptable risks, the risk treatment process is designed to plan for the implementation of Annex A controls and thereby reduce the impact and/or possibility of risks. At this point, a risk assessment report is to be documented at each phase of risk assessment and treatment. Additionally, it is necessary to get the go-ahead for any potential residual risks.

7) The SoA

You will know exactly which controls from ISO 27001 Annex A are required once you have finished the risk assessment and treatment procedure. The **Statement of Applicability** (**SoA**) is a document that details the organization's approach to determining which controls are relevant and which are not, along with an explanation of why some policies are deemed irrelevant and how others are applied.

ISO 27001 deems the SoA necessary. It effectively documents the ISMS control environment. It is based on risk assessments and risk treatments; therefore, it includes both implemented controls and risk treatment controls. Annex A is the template.

An SoA should address the following elements, as per clause *6.1.3 d* (as shown in *Figure 5.2*):

- The definition of which controls (security measures) will be applied, from the controls proposed by ISO 27001 Annex A

- The rationale for the inclusion of controls that are relevant to the situation

- The current implementation status of any controls that are applicable (that is, whether they are implemented or not)

- An explanation for the removal of controls from Annex A that are not relevant to the situation

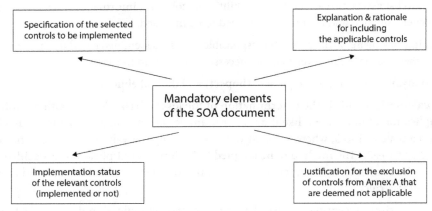

Figure 5.2 – Mandatory elements of the SoA

As a result, the SoA provides an all-encompassing perspective of the state of information security throughout the company. Also, the SoA is the best way to get management buy-in for introducing the ISMS. This is also the go-to document for the external auditor of the company.

After the SoA is prepared, the next step is to create a practical and solid risk treatment plan.

8) Risk treatment plan

The **risk treatment plan** is written after the SoA. The risk treatment plan should specify who will be in charge of implementing the SoA's controls, how much budget will be allocated and when, and so on. Your controls are the primary emphasis of this document, which serves as an action or implementation plan.

The following elements are required in a risk treatment plan:

- What the security controls and other activities to implement are
- Who is responsible for the implementation
- What the timelines are
- Which resources (i.e., financial and human) are required for the implementation
- How to determine whether the implementation was done correctly

In conclusion, a thorough SoA and an effective risk assessment and treatment procedure will provide the groundwork, but the risk treatment plan is where the execution starts.

Next, we implement the security controls and thereby enhance the security processes.

9) Implementing the security controls

Here is where you'll need to put in place all the necessary documentation and technology, thereby altering how your business handles security.

The following is a (non-exhaustive) list of documents essential for ISMS implementation:

- The ISMS scope document (clause 4.3)

- The information security policy, and the organization's context and objectives (clauses 5.2 and 6.2)

- The methodology for risk assessment and risk treatment (clause 6.1.2)

- The SoA document (clause 6.1.3 d)

- The risk treatment plan (clauses 6.1.3 e, 6.2, and 8.3)

- A report of risk assessment (clauses 8.2 and 8.3)

- Defined security roles and responsibilities (controls A.6.2 and A.6.6)

- Asset inventory (control A.5.9)

- Policy on the acceptable use of assets (clause A.5.10)

- Policy for access control (control A.5.15)

- Documented operating procedures (control A.5.37)

- Principles for secure system engineering (control A.8.27)

- Policy for supplier security (control A.5.19)

- Procedures for incident management (control A.5.26)

- Procedures for business continuity (control A.5.30)

- Legal, regulatory, and contractual requirements (control A.5.31)

The following are the mandatory records:

- Training, skills, experience, and qualifications documents (clause 7.2)

- Results from monitoring and measurement (clause 9.1)

- Internal audit program details (clause 9.2)

- Internal audit results (clause 9.2)

- Proof of conducting management reviews (clause 9.3)

- Corrective action results (clause 10.1)
- Log records of user activities, exceptions, and security events (clauses A.12.4.1 and A.12.4.3)

In addition to those in the preceding lists, there are several non-mandatory documents ranging from documents for control procedures, internal audit procedures, and **Corrective Action** (**CA**) procedures to policies for **Bring Your Own Device** (**BYOD**), mobile devices and teleworking, information classification, and so on.

The most difficult part is usually enforcing new procedures and methodologies in the company. Employees will always be wary of change; hence, the key is to empower them by providing education and training.

10) Implementing training and awareness programs

Explaining the rationale behind the new policies and procedures, as well as providing the appropriate training, is essential if you want your staff to follow them. The second most common cause of ISO 27001 project failure is insufficient training and awareness.

All employees must possess the necessary skills to perform their assigned duties effectively within the ISMS. Any identified skill gaps must be filled. However, certain groups receive training that is tailored specifically to the ISMS. The following list describes some of these communities and the possible training that may be necessary.

When you are developing any training plan, the following must be considered:

- The target audience
- The training contents
- The mode of training delivery
- The time and frequency of the training
- The personnel responsible for organizing the training, updating the content, and its delivery
- The requirement for assessments or evaluations

A training program can be created after data has been gathered.

The implementation of an ISMS depends heavily on effective communication. As a result of effective communication campaigns, data security is no longer neglected as a *side problem*. When it comes to the day-to-day operations of an ISMS, good communication approaches go far beyond deployment. A well-organized system of security dashboards, briefings, and warnings helps to keep everyone in the loop.

When it comes to effective communication in the workplace, you need to know your audience, the methods you can employ (either existing or new), the substance of your communications, and how often you should be sending them. When corporate communications departments are involved, the benefits of this project are enormous.

11) Operationalizing the ISMS

At this point, ISO 27001 should be ingrained into your company's daily operations. There are references to the needed paperwork in *clauses 4* through *10*. Both records and documents may be included in this documentation. Records serve as proof of a particular action across time.

Records are created (manually or automatically) when a specific activity is carried out, and they reflect what has been accomplished. For instance, your backup system will generate logs if your backups are carried out automatically. You can keep track of events with the aid of records. If there is a visitors' book, logging visitors' names and information is an example of a record. Records are therefore the evidence demonstrating that a task was completed.

Periodically, documents are updated and evaluated. Usually, they are versioned. To make sure that processes are carried out in line with the management system's goals, documentation is crucial. Documentation outlines your plan of action and offers proof that you are carrying out your promises.

Table 5.1 lists some examples of documents and records:

Name	Clause	Documentation type
ISMS scope	4.3	Document
High-level information security policy	5.2	Document
Risk assessment methodology	6.1.2	Document
Risk assessment report and risk treatments	6.1.2, 6.1.3, 8.2, and 8.3	Record
Statement of applicability	6.1.3 d)	Document
Information security objectives	6.2	Document
Evidence of competencies	7.2	Record
Monitoring and measurement results	9.1	Record
Internal audit program and results	9.2	Record
Results of management review	9.3	Record
Non-conformances and results of CA	10.1	Record

Table 5.1 – Examples of documents and records

The standard does not specify the amount of documentation required; instead, it is determined by a variety of factors, including the organization's size and primary functions, the complexity and interdependence of business processes, the control environment, the level of expertise of the workforce, and any legal or regulatory requirements. The target audience should always be kept in mind when writing a document. People who access the information in the documentation should find it beneficial. It is expected that organizations define and document their processes.

It shouldn't take a significant step to transition the ISMS from an implementation project to an operating system. Utilizing artifacts and processes as soon as they are created is key to a seamless transition. The operational areas should start using a process as soon as it is created. This strategy has many advantages. First off, the company does not consider this a "big-bang" strategy. Early recognition of the advantages of these new or modified processes strengthens support for the ISMS. The early collection of data and evidence to support ISMS improvement yields another significant benefit. It is crucial to make sure that the operational teams have access to these new and modified processes. The ISMS's operations should become *business as usual*.

There are certain internal hurdles to transformation in many organizations. The effective implementation of an ISMS may be hampered by several obstacles. These obstacles could be the company's culture of resistance to change, the fact that the organization is large and changes slowly, the requirement to be risk-averse due to being in a particular industry, employees feeling under-resourced and overburdened, and so on. This is especially pertinent when the implementation is motivated by a compliance objective. Effective and consistent communication methods that emphasize the advantages of the ISMS and *what's in it for them* are necessary to overcome these barriers. Another element that can aid in overcoming some of these difficulties is keeping the system as straightforward as possible.

A security calendar may be used to monitor the scheduled tasks in the information security domain. Even though it is not specifically required by the standard, a security calendar can be helpful in planning the execution of routine or scheduled actions linked to either *clauses 4–10* or controls included in the SoA for an ISMS. It can then be used as a resource for planning purposes and as part of system governance and monitoring to ensure that planned activities occur as intended.

The ISMS security calendar could be part of a wider compliance calendar. Activities that can be scheduled include internal audits of the ISMS, reviews of privileged users, management reviews, reviews of risk registers, and annual security awareness training.

12) Monitoring and measuring the ISMS

As popularly said, only what can be measured can be managed.

Monitoring an ISMS can be done in several different ways. One of the most important techniques for evaluating the performance of an ISMS and driving changes is measuring the effectiveness of the many components that make up the ISMS. It is up to the organization to decide what metrics are important, as well as how they will be measured and published.

For example, consider the case of backups in an organization. A company sets its objective to tolerate data loss for a maximum of five hours. This is a SMART objective and can be checked against how much data can be restored.

The ISMS certification process and the actual certificate both align with a specific version of the SoA. Therefore, if the number of controls changes, the certified business should construct a new SOA and ask for a new certificate. Although, note that **Certification Bodies (CBs)** may charge for the issue of a new certificate if it is outside the standard recertification cycle. Since the SoA is a byproduct of the risk assessment/risk treatment process, it is frequently associated with the risk register. When preparing an audit and selecting the audit team, the SoA is essential for enabling the development of the audit plan and guaranteeing the appropriate audit resources are selected. An ISMS built on the ISO 27001 standard must comply with *clauses 4* through *10*.

The SoA specifies those controls that are applicable to the organization out of the 93 controls listed in Annex A and why.

It's important to keep in mind that management systems with more experience often have access to more metrics than systems with less experience. More attention should be given to creating a reliable metric the more crucial or intricate the control area. A common strategy is to select controls depending on the kinds or degrees of risks they reduce. A control is more likely to be relevant and should be measured if it mitigates a larger number of risks. Measuring the efficacy of risk controls with higher ratings is another method. The company makes the choice in this regard.

Regardless of how you plan to measure the success of your security processes and/or controls, make sure you have a clear description of how you intend to measure your overall ISMS and the security processes and/or controls.

13) Internal audit

An ISMS internal audit mainly has two objectives (clause 9.2). The first is to determine whether the ISMS complies with the standard's requirements, the organization's internal policies and practices, and the legal and regulatory environment in which it works. The report generated as a result of this phase of the ISMS audit includes statements of conformance and non-conformance with those criteria.

Identifying improvements to the ISMS is the second purpose of the ISMS Internal Audit. The ISMS audit program serves as the organizational umbrella under which internal audits are carried out. This program often extends over the course of several years and provides an overview of the scope for each of the anticipated audits that is included in the program. Audits need to take place at regular planned intervals. It is necessary for the audit program to cover all mandatory clauses as well as all controls that are outlined in the SoA.

It's possible that each individual ISMS audit will only focus on a subset of the system's control domains and clauses. Without prior authorization, the auditor for each of these audits is not permitted to conduct audits that go beyond the purview of the respective audit. Audits of an ISMS always put an emphasis on the system rather than the individuals being audited. If any resource flaws are discovered, they must always be linked to a system weakness. It's possible that the mistakes were made due to a lack of understanding on the auditee's part of the obligations they have, a gap in their competency, or poor supporting policies and procedures. These are the issues that call for immediate attention and correction. An ISMS audit is more than just an evaluation of the controls. The management system is the most important component. When controls fail, it almost always indicates that one of the essential ISMS components has also failed. In most cases, control failures can be eliminated simply by fixing the underlying problem.

ISMS auditors, like all other jobs within the ISMS, are required to be knowledgeable and possess the requisite abilities to carry out an ISMS audit. While a general ICT auditor will have the knowledge to examine the controls, knowledge of the management system's workings is absolutely essential.

In a nutshell, internal audits can help you find issues (that is, non-conformities) that could be detrimental to your organization.

14) Management review

The management review's goal is to evaluate how well the ISMS is functioning considering a number of factors and to make adjustments as needed (clause 9.3). Senior leadership regularly performs Management Reviews at predetermined times. This is the typical ISMS governance meeting.

According to ISO 27001, management reviews must consider the following required inputs:

- The update on actions taken because of prior management evaluations
- Changes in ISMS-related external and internal issues
- Monitoring and measurement outcomes, audit findings, the accomplishment of information security goals, and feedback on performance
- Feedback from interested parties
- Risk assessment results and the status of the risk treatment plan
- Opportunities for continual improvement

The Management Review will produce decisions on opportunities for continual improvement and any adjustments to the ISMS. The findings of these procedures must be documented, often in the meeting minutes from the Management Review.

The ISMS Internal Audit should precede the Management Review since its findings form an integral element of the review.

15) Correction, corrective action, and improvement

The goal of a management system is to ensure that non-conformities are addressed or, ideally, avoided. As a result, ISO 27001 demands that corrections and CAs are carried out in a systematic manner, which means that the fundamental cause of non-conformity must be identified, remedied, and confirmed. ISO 27001 mandates organizations constantly upgrade their ISMS. CA is one strategy for driving system improvements and addressing system weaknesses. When non-conformity is discovered, ISO 27001 requires correction and further CA(s).

The need for CA can be the result of internal audits, external audits, management reviews, security incidents, or security reviews and testing. A risk owner can oversee the process and check its alignment with the risk register.

Correction is the immediate action needed to eliminate the identified non-conformity.

An example of CA is as follows. An intruder was successful in entering your company building, surpassing the access control at the entrance. On investigation, it was found that the access control device at the entrance was not working. The correction would be the immediate deployment of a security guard at the entrance who will check incoming employees' access.

Corrective Action (CA) refers to the actions performed to eliminate the fundamental cause (root cause) of a process of non-conformity and prevent its future occurrences. The CA consists of responding to a process problem, putting it under control through containment processes, and then taking the required steps to prevent it from happening again.

CA procedures capture the organization's response to a requirement for CA. This protocol includes a root-cause analysis required to ensure that the non-conformance does not reoccur. Except for non-conformity on the part of certifying authorities, the timing and implementation of CA is at the discretion of the organization. CA for any non-conformity discovered during certification or surveillance audits must be implemented within specific periods.

An example of CA in the context of the previous scenario is taking measures to rectify the issue and finally make the device function.

Improvements are when companies who already have an ISMS that is operational are required to continue to work toward making their management system better. This is an essential component of all management systems, and an ISMS is no exception.

Improvements can be from sources such as internal audits, external audits, management reviews, security incidents, security reviews, and testing or suggestions from interested parties.

Though they needn't necessarily be implemented, suggested improvements should be considered. The organization chooses the improvements it believes will benefit the ISMS. Although suggestions from both internal and external auditors should be taken into consideration, implementation is not mandatory.

The organization establishes deadlines for putting agreed-upon improvements into action.

16) Certification of an ISMS

An independent, credible third party must assess an ISMS' compliance with ISO 27001, the organization's own policies and processes, and the legal and regulatory context in which the business operates before it can be fully certified.

The purposes of a certification audit are as follows:

- Proving compliance with ISO 27001 to clients and other stakeholders
- Minimizing the information security risks to clients and suppliers
- Instilling more trust in the organization's information security among consumers, suppliers, regulators, and other stakeholders

Most accredited CBs will offer a gap analysis or pre-audit before the formal commencement of the audit process. The certification process is divided into two stages – Stage 1 and Stage 2. First, the organization submits an application to a CB of its choice as the first stage in the certification process. An initial certification audit (Stage 1) is the following stage. This is followed by the final Stage 2 audit. If there is no significant non-conformity discovered following this Stage 2 audit, certification will be recommended.

> **Certification audit**
>
> The auditor examines the system's documentation during the Stage 1 certification audit to make sure it complies with the standard's requirements. This review concentrates on the *intent*, or whether the organization has implemented or plans to implement this control, from the perspective of controls. The evaluation is based on the SoA's controls. All of the required documentation specified by the standard will be examined by the auditor as well. The auditor searches for proof that the processes and controls have been implemented and are operating during a Stage 2 audit.

We have covered how the implementation is done in different phases for an ISMS. Now let us see the hours and efforts that need to be put in by personnel in various roles for the implementation.

Time, effort, and roles in an ISO 27001 implementation

It can take a few months for smaller businesses and up to a year or more for larger firms to implement an ISMS. A successful and fruitful rollout can improve operational metrics such as efficiency, effectiveness, and cost savings, and reduces the frequency with which events occur. In smaller businesses, the project manager will also serve as the security officer, but in larger businesses, the roles will be distinct. A professional project manager will oversee the project while a second person serving as the security officer will oversee overall security and take part in it.

ISO 27001 does not require creating a project team; however, doing so will be beneficial for businesses with 200 or more employees. For smaller businesses, having just a project manager who will manage the project alongside other team members will suffice.

Irrespective of the size of the company, it is good practice to include part of your employees in activities such as risk assessment, risk treatment, reviewing policies and procedures to ensure the alignment of security documents with existing business processes, and approving security goals and documents to ensure commitment and adherence to the business strategy.

Summary

The process of implementing an ISMS may appear daunting, but it is rather simple. Formally identifying and managing threats to the organization's information is what is at the core of an ISMS. As soon as the scope is clearly specified, it's easy to move on to the implementation process. As with any huge project, well-thought-out plans lead to a successful outcome. It is important to ensure that the documentation is *fit for purpose* and targeted at the intended audience. It is important to keep in mind that clauses 4-10 of the ISMS are necessary. These clauses demand that you handle requirements in each of them.

In the next chapter, we will see the incident management process of an ISMS implementation.

6

Information Security Incident Management

It is practically impossible for any organization to be able to work without any incidents. This is because neither people nor systems and technologies are perfect. **Information security incident management** refers to the steps taken to identify, manage, record, and evaluate security incidents and threats associated with information security. In an information technology infrastructure, this is a highly crucial step to take either after or before a cyber disaster takes place.

In this chapter, we will look at the entire information security incident management process, starting with what a security incident is and moving on to the step-by-step process of incident management. This will be followed by an evaluation of the effectiveness of the process by implementing the appropriate controls. We will also look into how incident management is formed in an organization and the related standards to look into.

Security incidents are inevitable and can impact the business health and brand reputation of an organization. Having a good incident management process can mitigate the effects to a great extent. Detecting and responding to issues early reduces financial and operational damage.

In this chapter, we will look at the following main topics:

- Information security incidents and incident management
- Information security events, incidents, and breaches
- The incident management step-by-step process
- Assessing the effectiveness of the process
- The incident management team
- Incident management-related ISO standards

Understanding security incidents and incident management

Figure 6.1 shows the representation of an occurrence of an incident. **ISO 27000** defines a security incident as *"a single or a series of unwanted or unexpected information security events that have a significant probability of compromising business operations and threatening information security"* (https://www.iso.org/). There can be different sources of incidents, such as employees, clients, and third-party vendors. When an incident occurs or a weakness is discovered in a system or service, a mechanism must be established to ensure that an organization can respond quickly and effectively. The first thing that needs to be done to accomplish this goal is to devise a plan to handle any security problems that may arise.

ISO 27035 is the standard that talks in detail about information security incident management. Information security incidents and vulnerabilities can be identified, documented, assessed, responded to, managed, and used to drive future efforts to strengthen security. This standard offers a framework for doing so. It requires an organization to have a well-defined incident management policy. The policy is a high-level document that serves as the foundation for lower-level procedures. The information security incident response plan should be the primary procedural document for incident management since it contains comprehensive procedural and technical information on an incident management strategy.

The negative effects of cyber damage can be mitigated with effective incident management, which can also prevent a cyberattack from taking place. A business that does not have an effective incident response strategy runs the risk of falling victim to a cyberattack, which might result in the data of the organization being compromised to a significant degree.

Figure 6.1 – The occurence of an information security incident

Terms and definitions

The following are some of the important terms and their definitions in the context of incident management:

- **Information security forensics**: The process of capturing, recording, and analyzing occurrences related to information security incidents, through the application of investigative and analytical methodologies and the preservation of the evidence, in alignment with the applicable legal and regulatory requirements.

- **Information Security Incident Response Team (ISIRT)**: A team comprising employees of a company who are both trusted and have the necessary skills to handle information security incidents. The organization may temporarily onboard external experts to investigate/respond to computer incidents.

- **Information security event**: A discovered system, service, or network condition suggesting a probable information security policy breach, control failure, or an unforeseen security-relevant circumstance.

- **Information security incident**: A confirmed or suspected violation, compromise, or disruption of the confidentiality, integrity, or availability of information or information systems. It involves an actual or potential adverse event that poses a threat to the security of an organization's data, systems, or network.

- **Information security breach**: The unauthorized access, acquisition, disclosure, or loss of sensitive or confidential information. It signifies a specific incident where protected data is accessed, disclosed, or used by unauthorized individuals or entities in a manner that violates legal, regulatory, or policy requirements.

The benefits of implementing an incident management system

- Improves and strengthens the information security posture of an organization

- Helps to reduce catastrophe in the face of a security attack

- Strengthens and streamlines the information security incident prevention, prioritization, and evidence collection processes

- Helpful to plan for budget and resource allocations

- Improves the information security risk assessment and management processes of an organization

- The ability to provide enhanced and up-to-date information security awareness and training program material

- The ability to improve the information security policy of an organization and the related documentation

In the next section, we will see what security incidents and security breaches are and whether they are one and the same.

Information security incidents and breaches

If a company's security policy is violated, it results in a **security incident**. It can be an event that compromises any one pillar of the **Confidentiality, Integrity, and Availability (CIA)** triad. A **security breach** is when an unauthorized entity gains access to the organization's data, network, applications, or devices, which results in the disclosure of critical/sensitive information. An incident may or may not evolve as a breach.

Let us investigate a few examples to understand the difference between a security event, incident, and breach better:

- Let's imagine that in the building of organization *XYZ*, a window that provides access to physical files with personally identifiable information is accidentally left open. This is an **event**. Now, if a couple of files are missing, resulting from this careless act, it results in an **incident**. If someone with malicious intentions gains access to the files and, as a result, the organization's critical/ sensitive data is stolen, that becomes a **breach**.

- In another example, an employee leaves critical information written out on paper, unattended, at their desk. This is an event. Another employee passing by notices the unattended paper with critical information and takes a quick glance, but they don't make a deliberate attempt to retain or misuse the information. This escalates the situation to an incident because unauthorized access to sensitive information has occurred. The organization should investigate, document the incident, and take appropriate action to prevent similar incidents in the future. In a more severe scenario, a malicious individual seizes the opportunity when the critical information is left unattended. They intentionally take a photo of the paper or copy information from it without authorization. This unauthorized access and acquisition of sensitive information constitutes a breach. The perpetrator now possesses the critical information, potentially leading to misuse, unauthorized disclosures, or further security compromises.

- Let's suppose a user receives a suspicious email with a link to a potentially harmful website. This event raises a red flag but does not confirm any actual security incident or breach. After receiving the suspicious email, the user clicks on the link, unknowingly triggering the execution of malware on their computer. This action leads to unauthorized access to the user's personal files and sensitive information. The incident occurs when the user's system is compromised and their data is at risk. Now, the user, unaware of the email's malicious nature, enters their login credentials on the website linked in the email. The website, designed to mimic a legitimate service, captures the user's login information. The breach occurs when the hacker successfully obtains the user's login credentials and gains unauthorized access to their account or other sensitive systems. This unauthorized access exposes the user's personal data and potentially puts other users' information at risk.

There are several different techniques to determine whether your organization is at risk of experiencing a severe breach in its security. Different types of breaches will leave different traces, or markers, behind. Considering several factors, such as the ones listed as follows, can help organizations equip themselves in the face of an incident:

- Be on the lookout for anything out of the ordinary, such as suspicious file downloads or uploads, excessive use of data, or unauthorized account access attempts. A business that notices a sudden spike in traffic should investigate the source and be prepared for an attack.

- Check for attempts to access systems without permission and the overuse of resources (such as memory, hard drives, and networks).

- Monitor employees' access to outside information sources, as most of the time employees are the entry points for an attack.

- Verify the transfer of overly large files that could potentially hold payloads to compromise a system.

Information security breaches have the potential to put an organization's business systems at risk and disrupt normal operations.

An information security incident management system enables organizations to respond to incidents quickly and effectively, by implementing controls and procedures to curb information security incidents and manage vulnerabilities.

Security Incident Event Management (**SIEM**) is a powerful tool utilized by many organizations to enhance their security posture. It enables the collection, analysis, and correlation of security event logs and network traffic data from various sources within an organization's infrastructure. SIEM provides real-time monitoring, threat detection, and incident response capabilities, allowing organizations to identify and respond to security incidents more effectively.

In the next section, let us see step by step what the actual incident management process looks like.

The incident management process

Preparedness is essential for being effective in the event of a significant incident. This is a common-sense statement, yet it is not always followed in practice. In most cases, only after a few major incidents have occurred it is common for an organization to develop a set of incident-handling procedures, testing and adjusting those processes to meet its needs. Some organizations only equip themselves to handle an issue in part, and a comprehensive system that can deal with any form of an incident may not be present.

The first step is to find out which security events should be investigated, and at what thresholds, by also considering the business continuity requirements of the organization. The next step is to draft a response strategy for each different kind of incident. It is possible to improve it through security event simulations, which allow you to uncover gaps in your process, but it will also be improved after actual events have occurred. Develop a communication strategy that includes instructions on who to contact, how to reach them, and when to contact them, based on the type of situation.

Consequently, it is essential to have in place the following:

- A standard operating procedure or set of guidelines for handling common incidents
- A procedure to establish and assess such policies and procedures, as well as a unified plan to respond to and recover from any issues that may arise
- Methods to decide what to do next
- Establishing channels of communication and assigning roles during an incident

Prior to any possible crisis, it is essential that these procedures are tested thoroughly.

The planning exercise should develop a team with clear roles and duties to respond to incidents. This preparation exercise must also designate who to communicate with in the event of an incident – internal stakeholders (executive management, security, IT, legal, PR, HR, etc.) and external stakeholders (vendors, internet service providers, etc.).

A *four-step process* for incident response/management in information security is depicted in *Figure 6.2.*

Figure 6.2 – The incident management process

Detection and reporting

The incident management process begins with the detection of an incident that may potentially harm a business. The person who detects the potentially harmful event communicates according to the defined procedures in the company (via email, a phone call, a software tool, and so on).

This is where you go into research mode. Investigate the occurrence. Analyze. Locate, contain, and measure the breach. Having all your security tools in one place simplifies and speeds up this procedure.

A well-established system is in place for users to report security incidents, such as a dedicated email ID or a hotline number. The continuous monitoring of critical resources by the **Security Operations Center (SOC)** will also contribute to the timely detection of abnormalities.

Analysis and classification

The incident response team needs to review and evaluate the information that is reported to determine the magnitude and severity of an incident. A few questions and considerations here to arrive at key conclusions are as follows:

- Was this a legitimate security breach or some other event?
- The information should be checked for any inaccuracies and false positives or negatives.
- It's important to know the extent of the problem. Indicating which aspects of a business are affected should be part of the scope.
- The origin and cause of the problem – root-cause analysis.
- Is the incident contained or spreading? If it is spreading, then how fast?

The incident is categorized based on a variety of factors and considerations. Rather than the individual who discovers the incident, a technical expert is expected to classify it in the most accurate manner.

Incidents can be categorized in a number of ways; however, two factors are looked at mostly:

- **Impact**: The loss suffered by a business (financial, brand reputation, and so on)
- **Urgency**: How quickly an organization needs to fix the problem

The interplay of these two factors will decide each incident's priority, as shown in *Table 6.1*.

Impact / Urgency	High	Medium	Low
High	1	2	3
Medium	2	3	4
Low	3	4	5

Table 6.1 – The priority matrix

The incidents with a value of *1* are *urgent* and have a big impact, and hence, they need to be fixed first.

All stakeholders, both inside and outside a company, need to be informed about the occurrence of the incident. The incident response team needs to do a thorough investigation, verification, and documentation process for every incident. A preliminary investigation needs to be done to figure out the extent of an incident and gather enough information to rank the actions to be taken and decide the kind of response strategy that should be used (e.g., eradication, containment, and so on).

Treatment

After categorizing the incident and agreeing on the severity and time needed to fix it, a technical expert must decide what steps need to be taken to fix it. Assess the incidents to figure out what steps to take next to reduce the risk. There are three parts to the treatment process – containment, eradication, and recovery.

The goal of containing is to stop the spread. This is where you stop the threat from escalating. The goal of eradication is to eliminate the threat. You'll need to do more if the threat entered one system and spread to others. Recovery restores the system or returns it to normal if it didn't.

The outcomes from the treatment process are as follows:

- Restoring functionality and connectivity to computer networks and other affected IT infrastructure. Reconfiguring systems and networks and recovering lost or corrupted data is also part of the recovery process, as are applying patches and updating software, as well as resetting user credentials and passwords.

- Verifying the full functionality of all company systems, processes, and services, and communicating this to appropriate stakeholders, including employees, managers, customers, and suppliers.

- A full recording and documentation of the incident and any necessary adjustments to the system, rules, procedures, and processes, with final actions taken to end the incident case file.

Correction and corrective action

Correction is an action taken to address a non-conformity or correct an existing problem. It may be carried out by changing the way something is done or made. The aim here is to immediately fix the issue. **Corrective action** is the action taken to eliminate the cause of a non-conformity and to prevent it from happening again. It starts with identifying the root cause of the problem and then taking the required steps to eliminate it. The aim is to rectify systemic issues.

For example, an employee could click on a phishing link received via their email, unaware that it is fraudulent. Let's assume that the result was a breach of credentials (the employee's username and password). In this scenario, the correction steps would be to block the phishing URL at the organization's firewall, change the credentials (reset the password), and train the employee on phishing attacks. The company can adopt corrective action by intensifying the security training given to its entire staff, as it may identify a lack of awareness of the root cause of this issue. This step enables the company to equip itself against future incidents.

Post-incident activity

Record what you learn from every incident. All the information gathered during the incident's investigation is essential to preventing future incidents as well as to keeping evidence. Imagine a scenario in which a user unknowingly clicked on a phishing link received via email, following which a system was infected and exhibited abnormal behavior. The user then logs an incident, which is subsequently rectified and closed. All of the data collected to resolve the incident is stored in a knowledge base so that if the same issue arises again, the most appropriate solution can be accessed.

This portion of the procedure is essential for two reasons – first, to conduct a review of the incident situation, and second, to implement the appropriate preventive measures to safeguard an organization from future incidents. Along with producing an incident report, a few key steps that should be conducted under post-incident activities are as follows:

- Evaluate the efficiency of the incident handling procedure and the controls in place to protect information. Identify improvements in the incident management process.
- Identify the scope of improvements to the **Information Security Management Systems** (**ISMS**).
- Improve the policies, procedures, processes, and training at an organization, as necessary.
- Update contracts and SLAs if necessary.
- Update the risk register where necessary.

In *ISO 27001*, Annex A, the *A.5.24* to *A.5.28* and *A.6.8* talk about the life cycle of information security incidents. Let us investigate each control in detail.

Understanding the controls related to incident management from Annex of ISO 27001:2022

Security incidents, events, and weaknesses are addressed in the **Annex A** controls from *5.24* to *5.28* and *6.8* of ISO 27001. These controls aim to ensure that incidents, events, and weaknesses are handled in a uniform and effective manner throughout their life cycle.

A.5.24 – information security incident management planning and preparation

To provide a prompt, efficient, and well-organized reaction to resolve vulnerabilities, events, and security incidents, the management of an organization needs to define roles, responsibilities, and procedures. An incident is where the CIA triad is affected.

Prior to the occurrence of an incident, the protocols for incident, event, and response planning need to be well established and approved by the leadership. During an audit, the formal, documented procedures are expected to be in place, along with evidence that they are working.

A.6.8 – information security event reporting

Security incidents should be reported to the proper management channels at the earliest opportunity. In the event of an information security incident, all parties involved should be made aware of their reporting duties, reporting protocols, and points of contact.

A section on reporting security issues must be included in the regular training for all employees and anyone who may be affected by it, such as suppliers. Ideally, you should be aware of what defines an information security weakness, occurrence, or incident to execute this properly. In the event of an information security incident, the nominated information security administrator must be notified immediately, and the incident must be documented accordingly.

An auditor may randomly sample the general staff to check for their awareness of what a weakness, event, or incident is, how to report them, and what their responsibilities are.

All parties who use an organization's information systems and services are obliged to report any flaws and incidents that they observe or suspect. The reporting technique and mechanism should be easily available so that parties can promptly disclose weaknesses to the designated point of contact to prevent incidents. Employees must also understand that when they uncover a security flaw, they must not seek to prove it, as doing so could be construed as a misuse of the system and could result in security incidents.

A.5.25 – assessment and decision on information security events

A thorough investigation is required before an event can be called an "incident" and deciding on an appropriate course of action. The impact and extent of an information security event must be assessed using an agreed-upon classification scale to determine whether it constitutes a security incident. Ideally, the process should have minimum impact on other users of the services. For future reference and verification, the outcomes of this evaluation must be documented.

A.5.26 – response to information security incidents

Actions taken in response to a breach in information security should follow predetermined procedures. There should be designated personnel in charge of handling any events related to information security.

Responsibility assignments, detailed job descriptions, and audit logs are all highly recommended practices in accordance with ISO 27001. When a security incident happens, the person in charge is tasked with bringing things back on track, all while gathering evidence, doing an information security forensics analysis, and alerting parties if required. In addition to reporting the presence of an information security event or other pertinent facts to leadership and other stakeholders on a need-to-know basis, the person in charge will also ensure that all response actions are appropriately recorded for further analysis.

A.5.27 – learning from information security incidents

All incident-related knowledge must be utilized to prevent future incidents. Information security incidents must be recorded, monitored, and analyzed. Through these analyses, reoccurring or high-impact accidents should be identified and prevented.

The IS policy has to demonstrate how lessons learned from investigating and resolving events will be applied to prevent or mitigate such occurrences in the future. Using the lessons learned from any security event to enhance services in the future is crucial to the ISMS's ability to grow and change with the times.

The incident's primary respondent can propose changes to ISMS policies, which the ISMS board can then consider. When the analysis and new knowledge have been absorbed, the appropriate personnel must be informed and retrained, and the information security education and awareness cycle will continue.

A.5.28 – collection of evidence

If any kind of legal proceedings is expected from an occurrence, an organization must create controls to identify, collect, acquire, and store evidence. Collection of evidence should be done meticulously if there is a chance of the security incident leading to legal or disciplinary action. Disciplinary procedures should also be clearly linked to information security incidents.

A solid incident management team is essential for the effective execution of the incident management framework set out by an organization. In the next section, we will see how an incident management team is constituted at various organizations.

Understanding the roles and responsibilities of the incident management team

An information security incident management process entails the steps necessary to finding, analyzing, and fixing issues, as well as speeding up the recovery of systems that have been compromised. An interdisciplinary team with a range of expertise is needed to address the myriad of challenges that may arise during this procedure. Management, IT and non-IT workers, physical security personnel, human resources, legal counsel, public relations specialists, reporters, and even emergency services may all be involved in the incident.

The incident response team manages cyberattacks, system failures, and data breaches. These teams may also create incident response plans, identify and fix system vulnerabilities, enforce security policies, and evaluate security best practices.

Incident management teams at various organizations may be referred to by different names, and except for a few differences, these teams essentially perform the same responsibilities. One example is the SOC, which covers a broader scope of cybersecurity. This includes incident response, as well as the monitoring and defense of systems, the configuration of control mechanisms, and the overall management of an organization's operations.

The team must be able to assess the situation, ascertain the extent of the breach, take the necessary steps to fix the issue and avoid a recurrence, and keep in touch with all the involved parties, both inside and outside an organization. A leader should be appointed to oversee the incident management team and be responsible for facilitating and managing the deployed team.

To put together an incident response team, the first step is to recruit people with the appropriate skill sets and backgrounds. The most efficient teams consist of a diverse collection of individuals that can assist in the management of all facets of an incident and offer a wide range of skills. The Incident Response Team (IRT)'s staff members should have backgrounds in cyber forensics, functional approaches, and maybe law and governance as well.

A typical incident response team and their roles and responsibilities are listed in *Table 6.2*.

Role	Responsibility
Team leader	Coordinates the incident response team and is responsible for keeping the team focused on activities to minimize damage and recover quickly
Lead investigator	Conducts the primary investigation by collecting and analyzing the evidence, determining the root cause, and guiding the other analysts in their efforts to implement a recovery process
Analysts and researchers	Carries out the incidence response activities and provides threat intelligence and context of the incident
Communications member in charge	Manages communication in the team and across the organization, ensuring that other relevant stakeholders such as customers and the public are informed about the incident
HR/legal representation	HR deals with employee involvement, and the legal team provides guidance regarding compliance and law enforcement

Table 6.2 – IRT members

Depending on the team's composition, it may be necessary for a single individual to perform two tasks, or for multiple people to concentrate on a single function. It is a challenge when companies span across multiple locations, as do most security incidents. Try to establish a local presence where most of the business and IT operations take place, even if a primary member of your team cannot be on site at every location. In many cases, you may need to have physical access to carry out investigations and analyses.

Once the team is ready, they can start preparing for and dealing with IT incidents. However, even with thorough planning, responding to incidents may be extremely stressful, especially for teams who lack prior experience. Centralized management simplifies incident response. Centralizing data makes analysis easier and boosts accuracy by providing context for analyzed events. SIEM solutions are the most typical technique to centralize data. These solutions aggregate data from multiple platforms.

Teams must also avoid gut reactions and investigate circumstances thoroughly. Act on objective evidence. Casually discarding an incident can later lead to more impactful attacks. Assuming what triggered an alert or how to fix it without first confirming your concerns can cause damage or result in an ineffective response. Investigating incidents thoroughly enables accurate incident identification. You can respond effectively and efficiently without wasting time or risking systems or workloads.

Establishing a team to respond to incidents with well-defined tasks is the key. Other support positions, such as those in finance, law, communication, and operations, will also be part of the IRT.

ISO/IEC 27035 is the major standard that describes information security incident management. Apart from that, there are a few other standards that also explain each step in the process. Let's look at those and an overview of what they stand for.

In the ISO family of standards, ISO 27035 covers the information security incident management part. However, there are also related standards that guide analysis and forensic activities. Let us understand what all related standards are available for information security incident management.

ISO/IEC standards related to information security incident management

ISO/IEC 27035 is the guideline standard for information security incident management. There are also six other standards that relate to information security incident management in one way or the other. Let's explore all seven standards as follows:

- **ISO/IEC 27035** (*Information technology – Security techniques – Information security incident management*) (https://www.iso.org/): Information security incident management is explained, focusing on detection, reporting, evaluation, response, and lessons learned, which comprises three parts (https://www.iso.org/):

  - **Part 1**: The principles of incident management

  - **Part 2**: Guidelines for planning and preparing for incident response

  - **Part 3**: Guidelines for ICT incident response operations

- **ISO/IEC 27037** (*Information technology – Security techniques – Guidelines for identification, collection, acquisition, and preservation of digital evidence*) (https://www.iso.org/): In this standard, you'll learn how to identify, gather, acquire, and preserve digital evidence that may be useful in a legal case.

- **ISO/IEC 27038** (*Information technology – Security techniques – Specification for digital redaction*) (`https://www.iso.org/`): This outlines the features of several methods to digitally redact documents. It also outlines the specifications for software redaction tools and ways to verify that digital redaction has been accomplished securely.

- **ISO/IEC 27041** (*Information technology – Security techniques – Guidance on assuring suitability and adequacy of incident investigative method*) (`https://www.iso.org/`): This gives guidelines on how to ensure that the methodologies and processes used to investigate information security events are "fit for purpose."

- **ISO/IEC 27042** (*Information technology – Security techniques – Guidelines for the analysis and interpretation of digital evidence*) (`https://www.iso.org/`): This includes advice on how to analyze and evaluate digital evidence in a way that solves continuity, validity, reproducibility, and repetition difficulties.

- **ISO/IEC 27043** (*Information technology – Security techniques – Incident investigation principles and processes*) (`https://www.iso.org/`): It gives instructions based on idealized models for typical incident investigation methods across a wide range of incident investigation situations, including digital data.

- **ISO/IEC 30121** (*Information technology – Governance of digital forensic risk framework*) (`https://www.iso.org/`): This offers a structure to govern bodies of organizations on how to effectively prepare an organization for digital investigations before they happen.

The ISO/IEC standards related to information security incident management enable organizations to effectively respond to and manage security incidents. The previously listed standards outline best practices and guidelines for incident management processes, ensuring a structured and coordinated approach. By adhering to these standards, organizations can enhance their incident response capabilities, minimize the impact of security incidents, and strengthen their overall information security posture.

Summary

To conclude, reducing recovery costs, liabilities, and damage to information systems all depend on having a solid incident management process in place. The financial and operational effects of incidents can be mitigated through early detection and rapid response. It is essential to have an information security incident response plan in place to guarantee that your firm is prepared to deal with all information security problems. This reduces information security attack costs and prevents further breaches.

The framework or process for incident management is executed through a strong incident management team. The process itself takes inputs from the various standards published by ISO that relate to incident management.

In the next chapter, we will discuss the case studies of ISO 27001 implementation that relate to risk management, the implementation of controls, ISMS development stages, and incident management.

Case Studies – Certification, SoA, and Incident Management

This chapter delves into a series of case studies centered around the implementation of an **Information Security Management System (ISMS)**, the ISO 27001 certification process, the creation of a **Statement of Applicability (SoA)**, and the management of information security incidents. These case studies revolve around a hypothetical organization named Titan Consulting Inc., a rapidly growing technology consulting firm operating in the IT industry.

Each case study will provide a comprehensive analysis of Titan Consulting Inc.'s journey toward securing its information assets. We will examine its initial motivations for pursuing ISO 27001 certification, the steps taken to implement the ISMS, and the successful outcomes it achieved.

Furthermore, we will explore the process of preparing an SoA specific to Titan Consulting Inc., outlining the scope of its ISMS and the controls chosen to mitigate identified risks. External auditors commonly rely on the SoA as a valuable point of reference, along with the risk assessment and **Risk Treatment Plans (RTPs)**. Thus, the SoA serves as a critical component of the ISO 27001 certification process, providing a transparent and structured view of the organization's security posture.

Lastly, we will delve into the realm of information security incident management, exploring how Titan Consulting Inc. addresses an incident in different steps. By understanding its approach, we can glean important lessons in mitigating and responding to information security incidents, minimizing their impact, and enhancing overall organizational resilience.

The following are the case studies presented in this chapter, which depict the concepts that we covered in previous chapters:

- Case study #1 – ISMS implementation and ISO 27001 certification
- Case study #2 – selection of controls and the SoA
- Case study #3 – information security incident management

Case study #1 – ISMS implementation and ISO 27001 certification

This case study explains how the organization Titan Consulting Inc. implemented an ISMS and prepared itself for ISO 27001 certification.

Company profile

Titan Consulting Inc. is a mid-sized consultancy company with about 1,000 employees, providing consultancy services in **strategic management**, **change management**, **human resources management**, and the **development of integrated management systems**.

The headquarters of the company are in an administrative building in the center of town A. The headquarters comprise the offices of the CEO and those of the respective managers. The company has four other offices in various locations.

Organization chart

Figure 7.1 depicts the organization's structure and the roles and responsibilities:

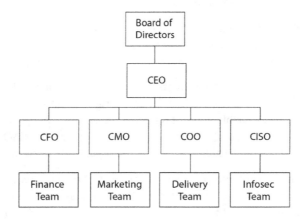

Figure 7.1 – Organization chart

An organization structure that includes a board of directors, CEO, CMO, COO, CISO, and other key roles is crucial as it establishes clear lines of authority, accountability, and decision-making within the company, ensuring effective governance and strategic direction. This hierarchical arrangement facilitates specialized expertise, efficient coordination, and effective risk management, enabling the organization to adapt, innovate, and thrive in a complex business environment:

- **Board of Directors**: The board of directors plays a vital role in overseeing the company's overall governance, setting strategic goals, and representing the interests of shareholders.

- **CEO**: The **chief executive officer** is responsible for the overall management and planning of the activities of Titan Consulting Inc., including resource planning and holding ISMS management reviews. The person occupying the position at present was promoted to this position by the board of directors after seven years in the company as an operations manager.

- **CFO**: The **chief financial officer** is responsible for the evaluation of investment projects, the management of assets, and the management of the accounting system, and participates in the ISMS management reviews.

- **CMO**: The **chief marketing officer** plans and implements the marketing strategy of the company and participates in the ISMS management reviews.

- **COO**: The **chief operations officer** is responsible for the day-to-day planning and management of the company's activities, general administration, and human resources and participates in the ISMS management reviews.

  The operations manager provides information security and operates, controls, and improves the ISMS.

- **CISO**: The **chief information security officer** oversees information security practices across the organization and ISMS management reviews.

Let us look at how Titan Consulting Inc. defined the scope and objectives for its ISMS, followed by establishing relevant policies and procedures.

Establishment and management of the ISMS

Establishing and effectively managing an ISMS requires a clear understanding of its scope, objectives, policies, and procedures. These components play a crucial role in ensuring the success and effectiveness of an organization's information security efforts.

Scope

The ISMS of Titan Consulting Inc. covers all the company's activities related to receiving, creating, processing, retaining, archiving, and transmitting information. The ISMS encompasses both in-house operations and interactions with third parties. Management's end goal in establishing and maintaining the system is to safeguard the company's information assets and its business continuity procedures.

The ISMS is established in compliance with ISO 27001:2022 and ISO 27002:2022, as well as with all applicable legal requirements.

Objectives

Setting clear objectives is vital for driving the organization's information security efforts. The objectives should align with the organization's overall goals and reflect the desired outcomes of the ISMS. The following are the set objectives:

- To handle information security according to corporate needs and legislation
- To attain and maintain a suitable level of asset protection
- To preserve the availability and accuracy of information and information processing facilities

Documents and records

Titan Consulting Inc.'s top management is committed to protecting the privacy of its customers and employees as well as the security of its financial data, client records, intellectual property, and other information assets in order to sustain and grow the company's market share, revenue, profits, and goodwill. The organization's ISMS was established based on a risk assessment using best practices, techniques, and information security experience.

The following are the key documents established by the company. This includes the mandatory ones for ISO 27001 certification:

- **ISMS scope document**: Outlines the boundaries and applicability of the organization's ISMS, including locations, departments, and functions to be included (clause 4.3).
- **Information security objectives**: Specific, measurable goals an organization sets to ensure the confidentiality, integrity, and availability of its information assets (clause 6.2).
- **Information security policy**: Sets the principles and guidelines for safeguarding and protecting an organization's information assets (clause 5.2).
- **User management policy**: Defines the processes and procedures for creating, managing, and revoking user accounts and access rights within an organization's information systems (controls 5.15, 5.18, and 8.3).
- **Physical/logical security policy**: Establishes measures and controls to protect physical and logical assets, including premises, equipment, networks, and systems, from unauthorized access, damage, or theft (controls 7.1, 7.2, 7.3, and 7.4).
- **Password policy**: Outlines requirements and best practices for creating and managing passwords to ensure strong authentication and protect against unauthorized access (controls A.5.16, A.5.17, and A.8.5).

- **Asset management policy**: Provides guidelines for the identification, tracking, and management of an organization's physical and digital assets throughout their life cycle (control A.5.10).

- **Email use policy**: Defines acceptable use and security guidelines for the organization's email systems to ensure confidentiality, integrity, and the proper use of electronic communications (control 5.14).

- **Information classification policy**: Establishes a framework for categorizing and labeling information based on its sensitivity and criticality and the regulatory requirements to facilitate appropriate handling and protection (controls A.5.10, A.5.12, and A.5.13).

- **Clear desk and clear screen policy**: Requires employees to keep their work areas tidy and free of sensitive information and mandates locking or logging off computers when unattended to prevent unauthorized access (control A.7.7).

- **Internet usage policy**: Outlines guidelines and restrictions for accessing the internet using the organization's network, specifying acceptable usage, prohibited activities, and measures to mitigate security risks (controls 8.20, 8.21, and 8.23).

- **Network security policy**: Defines the rules, configurations, and controls to protect the organization's network infrastructure from unauthorized access, malicious activities, and data breaches (control 8.20).

- **IT security policy**: Outlines the guidelines, rules, and procedures for accessing and using an organization's IT resources and information to ensure data integrity, confidentiality, and availability (control A.5.10).

- **Secure development policy**: A set of guidelines designed to ensure the secure design and development of software, preventing potential vulnerabilities and minimizing potential security risks (controls A.8.25 and A.8.27).

- **Remote work policy**: Sets guidelines and requirements for employees working remotely, including secure connectivity, data protection, and adherence to information security policies (control 6.7).

- **Antivirus and patch management policy**: Establishes procedures for deploying and maintaining up-to-date antivirus software and regular patches to mitigate vulnerabilities and protect against malware and security exploits.

- **Supplier security policy**: A set of guidelines and requirements that dictate the expectations and standards for information security practices that suppliers must adhere to when providing goods or services to an organization. It aims to ensure that suppliers maintain a secure environment for handling and protecting the organization's sensitive information throughout the supply chain. (controls A.5.19, A.5.21, A.5.22, and A.5.23).

- **Risk management process**: To identify, assess, and mitigate potential risks to the organization's information assets and operations (clause 6.1.2).

- **RTP**: A strategic document that outlines the approach and specific actions to be taken to manage, mitigate, or accept identified risks to an acceptable level (clauses 6.1.3 e, 6.2, and 8.3).

- **Statement of Applicability**: The key document in ISO 27001 that outlines which of the standard's controls are applicable and how they are implemented or justified for exclusion within the organization's ISMS (clause 6.1.3 d).

- **Risk assessment and treatment report**: A document that identifies, assesses, and proposes mitigation strategies for potential risks to an organization's ISMS (clauses 8.2 and 8.3).

- **Inventory of assets**: A detailed record of the organization's assets, including information, hardware, software, and personnel, critical for maintaining effective information security controls (control A.5.9).

- **Legal, regulatory, and contractual requirements**: Ensures the organization identifies, documents, and keeps up to date with laws, regulations, and contractual obligations related to information security and privacy (control A.5.31).

- **Agreements and NDAs**: To legally secure confidential information and define the terms of business relationships (controls A.6.2 and A.6.6).

- **Security incident management process**: To detect, respond to, and recover from security incidents, minimizing their impact on the organization's information security (control A.5.26).

- **Security operating procedures and configurations**: Documented and standardized protocols for system setup and operation, aiming to ensure consistent security practices and reduce vulnerabilities (controls A.5.37 and A.8.9).

- **Continual improvement process**: To assess and enhance the effectiveness of information security controls and processes over time (clause 10).

- **Information security awareness process**: To educate and raise awareness among employees about information security risks, best practices, and their responsibilities in safeguarding organizational information (clause 7.3, controls 6.3 and 8.7).

- **Password management process**: To ensure the secure management, storage, and use of passwords, reducing the risk of unauthorized access to systems and data.

- **Internal audit process**: To examine and evaluate the organization's information security controls and processes and ensure compliance with policies, standards, and regulatory requirements and identify areas for improvement (clause 9.2).

The following records are maintained by the company:

- **Training certificates and employee profiles**: To verify that candidates are selected in a transparent manner, based on their skills and suitability for the roles, and that the employees have received the necessary training to properly handle sensitive information and uphold information security standards (clause 7.2)

- **Monitoring and measurement report**: To ensure that the performance and effectiveness of the ISMS are continually tracked, allowing for improvements when needed (clause 9.1)

- **Internal audit program**: A systematic plan that facilitates the regular evaluation of an organization's systems and processes, ensuring they comply with internal and external standards, and identifying areas for improvement (clause 9.2)

- **Internal audit report**: Documents the detailed findings from internal audits, facilitating informed decisions on strengthening the ISMS and mitigating risks (clause 9.2)

- **Minutes of management reviews**: Documented evidence of management's commitment to the ISMS, ensuring continual improvement and adaptation to changes (clause 9.3)

- **Corrective action records**: Document actions taken to rectify non-conformities and prevent their recurrence, ensuring continuous improvement of the ISMS (clause 10.2)

- **Logs of activities and events**: Allows the identification, investigation, and mitigation of security incidents, aiding in the prompt resolution and prevention of potential security breaches (control A.8.15)

Titan Consulting Inc. has established a robust risk assessment process that systematically identifies and evaluates information security risks, enabling proactive measures to mitigate potential threats. The company has developed a comprehensive SoA and structured documented information as required by ISO 27001, providing a clear framework for implementing and maintaining an effective ISMS. With strong management commitment, Titan Consulting Inc. demonstrates a dedicated focus on the successful implementation and continuous improvement of its ISMS, ensuring the protection of valuable information assets and meeting the highest standards of information security.

Let us investigate these aspects in the following sections.

Risk assessment and treatment

To effectively safeguard its operations and protect sensitive information, Titan Consulting Inc. conducts comprehensive risk assessments to identify potential threats and vulnerabilities:

- **Risk assessment methodology**: The company has developed and implemented a method for assessing risk (based on the requirements of ISO 27001, 27005, and 31000), one that generates risk estimates based on the expected cost of damage caused by a given event and the likelihood of that event occurring. Once the risks have been evaluated, appropriate measures are made to eliminate them, lessen their impact, or pass them along to another entity. The "owners" of the assets have been determined by the application of a classification approach, which was accepted by management.

  Threats to assets, their vulnerabilities, and losses of confidentiality, integrity, and availability have also been documented.

- **Risk assessment**: The company has assessed the business harm from a security failure, including loss of confidentiality, integrity, or availability of assets. The likelihood of a security failure has been assessed based on existing threats, vulnerabilities, impacts, and controls.

  Risk levels have been estimated, along with tolerable and treatable hazards. The risk assessment report details the outcomes. Regular reviews and reassessments are planned.

- **Risk treatment**: The company has an RTP. Depending on the risk assessment, risk treatment actions, such as applying appropriate controls, risk acceptance, risk avoidance, or risk transfer, are taken.

The statement of Applicability

The SoA of Titan Consulting Inc. was adopted by top management at the information security forum held on XX-XX-2022 and will be reviewed for adequacy at least annually and when there are any significant changes to the ISMS. The statement considers the risk assessment held on XX-XX-2022. The identification of the controls selected followed that of *Annex A* of ISO 27001.

Management has implemented the selected controls according to the requirements of the standard, active legislation, and the contractual obligations of the company.

Documented information

Titan Consulting Inc.'s ISMS includes documented information required by ISO 27001 and that determined by the organization for the effectiveness of its ISMS.

All the documented information is properly structured with identification and a description (such as the title, date, author, and reference number), format specification (such as language and version), and sensitivity classification, and is reviewed appropriately.

Management commitment

Leadership demonstrates unwavering dedication to the ISMS's creation, rollout, operation, monitoring, review, maintenance, and improvement. Management has established information security goals and plans via the creation and adoption of an information security policy. Information security-related roles and responsibilities have also been defined. The information security policy, legal obligations, and the requirement for continuous improvement have all been communicated throughout the company. The administration allocates a sufficient budget to the ISMS's creation, rollout, operation, and management. Acceptable risks have been ranked and quantified. ISMS management reviews are scheduled at regular intervals or when there is a significant change to the ISMS that calls for a review.

To ensure the competence of employees in handling their assigned responsibilities as defined in the ISMS, an organization undertakes several crucial steps. These include determining the necessary competencies for a role, hiring skilled personnel, providing training, and evaluating the effectiveness of the training. Collectively, these measures contribute to the organization's ability to guarantee that employees are proficient in performing the required tasks. In order to accomplish the goals of the ISMS, the company ensures that all relevant employees fully grasp the significance of their information security operations.

Management review of the ISMS

To ensure the ISMS remains applicable, efficient, and effective, management conducts periodic reviews. The ISMS, comprising the security policy and the security objectives, will be reviewed to determine whether any changes are necessary. Data collected from the reviews is recorded and kept for future reference.

Management reviews are conducted to plan the enhancement of the ISMS by making changes to applicable policies and procedures, allocating necessary resources, and so on based on inputs such as the results of ISMS audits and reviews, the status of preventive and corrective actions, and follow-up actions from previous management reviews.

In order to ensure that the ISMS's controls, controls, processes, and procedures meet the criteria of the standard and any other applicable regulations, and to verify the efficacy of implementation, Titan Consulting Inc. prepares periodic internal ISMS audits.

The audited area's management promptly eliminates non-conformities and their causes, and checks and reports improvements.

ISMS improvements

The organization has committed to enhancing the effectiveness of the ISMS by using the information security policy, security objectives, audit findings, analysis of monitored events, corrective and preventative measures, and management reviews.

For the management and control of corrective actions, a procedure has been devised.

In this section, we saw how Titan Consulting Inc. has implemented an ISMS in the organization and prepared themselves to get ISO 27001-certified.

In the next section, we will see the controls Titan Consulting Inc. has decided to implement, based on which it has prepared the SoA.

Case study #2 – selection of controls and the Statement of Applicability

The SoA is an essential document that defines which controls from Annex A of ISO 27001 are being implemented in the organization and how, which controls will not be implemented, and the reason for their elimination (this reflects the risk appetite of the organization). When preparing an SoA, it is important to note that not all controls are mandatory for implementation. The selection of controls should be based on the organization's risk assessment, business requirements, and the specific context of the ISMS. This allows organizations to tailor the controls to their specific needs, ensuring a more efficient and effective implementation of the ISMS.

Tables 7.1, *7.2*, *7.3*, and *7.4* list the organizational, people, physical, and technological controls, respectively. Together, they form the SoA for the ISMS implemented at Titan Security Inc.

Organizational controls

ISO 27001:2022 incorporates controls focused on governance for organizations, termed organizational controls. The framework for administering and protecting the information assets of an organization is established by its rules, processes, and guidelines. These ensure that the company's security governance framework is well defined and that all stakeholders are aware of their specific responsibilities. Information security risk management also involves implementing rules and procedures and ensuring that they are regularly evaluated and updated.

The organizational controls defined in ISO/IEC 27001 (and elaborated in ISO/IEC 27002) are enumerated in *Table 7.1*, the controls applicable to Titan Security Inc. are indicated, justifications for including or omitting controls are provided, and a corresponding reference document is also provided:

Control Reference (https://www.iso.org/)	Control Name (https://www.iso.org/)	Applicable Yes/No	Justification	If Yes, Give Reference Document
5	Organizational controls			
5.1	Policies for information security	Yes	To define the organizational framework of the ISMS and set information security standards for the organization and external parties. These are reviewed and updated as business needs change.	• Information security policy. • Minutes of meetings of the information security committee meetings.
5.2	Information security roles and responsibilities	Yes	To ensure all requirements and controls have owners.	Information security policy.
5.3	Segregation of duties	Yes	To ensure accountability for each personnel's task and that there is no unauthorized access to other areas of duties.	Policy for segregation of duties.
5.4	Management responsibilities	Yes	The management mandates and ensures compliance with the information security policy and procedures on the part of all employees, contractors, and third parties.	Personnel security policy.
5.5	Contact with authorities	Yes	To ensure the cooperation of government organizations, healthcare companies, and other technology organizations.	List of contact details with administration department.

Control Reference (https://www.iso.org/)	Control Name (https://www.iso.org/)	Applicable Yes/No	Justification	If Yes, Give Reference Document
5.6	Contact with special interest groups	Yes	This is to obtain guidance and advice from information security specialists regarding information security.	IT team and administration department have the details.
5.7	Threat intelligence	Yes	To identify and mitigate threats.	Incident response and event management framework.
5.8	Information security in project management	Yes	To ensure that information is protected in all projects by including security objectives, security specifications, and risk assessment and making sure that security rules are included in all the tasks of the project.	Software project management procedure and infrastructure project management procedure.
5.9	Inventory of information and other associated assets	Yes	To ensure that all assets are accounted for by keeping an accurate inventory.	• Asset management policy. • Asset inventory.
5.10	Acceptable use of information and other associated assets	Yes	To ensure assets are used and maintained in accordance with the business objectives of the organization. This is to prevent any misuse of assets assigned to employees.	Acceptable use policy/ asset management policy.
5.11	Return of assets	Yes	Upon termination, employees are required to return all organizational assets in order to avoid data leakage and breaches of confidentiality.	• Employee clearance forms. • Asset management policy.

Control Reference (https://www.iso.org/)	Control Name (https://www.iso.org/)	Applicable Yes/No	Justification	If Yes, Give Reference Document
5.12	Classification of information	Yes	To ensure that assets are categorized according to the level of sensitivity and the appropriate level of protection is assigned to assets based on the classification.	• Asset classification guidelines. • Information classification policy.
5.13	Labeling of information	Yes	Based on the classification of an asset, the asset should be clearly labeled to ensure that assets are not protected and not mishandled.	• Information labeling and handling procedure. • Information classification policy.
5.14	Information transfer	Yes	To prevent information misuse by the recipient or through the medium of transfer.	Information exchange policy.
5.15	Access control	Yes	To ensure access to information assets and information processing facilities is approved, authorized, and controlled.	• Network and application access control policy and procedure. • Digital access management matrix. • Record of periodic review of the access control matrix. • LAN management process.
5.16	Identity management	Yes	To ensure the identification of individuals and systems accessing the organization's information.	Digital access management policy.

Control Reference (https://www.iso.org/)	Control Name (https://www.iso.org/)	Applicable Yes/No	Justification	If Yes, Give Reference Document
5.17	Authentication information	Yes	To ensure that access to information assets is only available to authorized users.	• Process for the issue and control of RSA tokens. • Password policy. • Digital access management policy.
5.18	Access rights	Yes	• To ensure the process of assigning and revoking access is established and maintained. • To ensure that user access rights are not misused and to review whether existing access rights have to be revoked or maintained. This prevents unauthorized access to information systems. • To ensure all access rights are revoked when an employee resigns or is terminated and that an employee does not have any unauthorized access upon change of employment.	• Digital access management policy. • Information security audit procedure. • Termination checklist.

Control Reference (https://www.iso.org/)	Control Name (https://www.iso.org/)	Applicable Yes/No	Justification	If Yes, Give Reference Document
5.19	Information security in supplier relationships	Yes	To define the minimum required security when providing access to third parties to the organization's systems.	Vendor management process.
5.20	Addressing information security within supplier agreements	Yes	To strike a balance between granting access to third parties and maintaining the necessary level of security, it is crucial to ensure that adequate security measures are implemented for the organization's systems.	Vendor management process.
5.21	Managing information security in the Information and Communication Technology (ICT) supply chain	Yes	To adequately address all risks associated with communication technology services in supplier interactions.	Vendor management process.
5.22	Monitoring, review, and change management of supplier services	Yes	Contractual requirement.	<ul><li>Third-party security compliance guidelines.</li><li>Record of management reviews of third-party services.</li></ul>

Control Reference (https://www.iso.org/)	Control Name (https://www.iso.org/)	Applicable Yes/No	Justification	If Yes, Give Reference Document
5.23	Information security for the use of cloud services	Yes	The organization's information security requirements should be the guiding principle for establishing processes related to the acquisition, utilization, management, and termination of cloud services.	Supplier security policy.
5.24	Information security incident management planning and preparation	Yes	For information security incident management and reporting.	Incident response and event management framework.
5.25	Assessment and decision on information security events	Yes	To assess the impact, evaluate the severity, and determine the necessity of applying the security incident response plan to address the incident.	Incident response and event management framework.
5.26	Response to information security incidents	Yes	To maintain adherence to the documented procedures of the security incident response plan, it is essential to ensure that security incidents are appropriately addressed.	Incident response and event management framework.

Control Reference (https://www.iso.org/)	Control Name (https://www.iso.org/)	Applicable Yes/No	Justification	If Yes, Give Reference Document
5.27	Learning from information security incidents	Yes	The organization is required to conduct reviews of security incidents and identify their root causes. The insights gained from these reviews will be utilized to strengthen the system and minimize both vulnerabilities and the frequency of future incidents.	• Incident response and event management framework. • Incident register.
5.28	Collection of evidence	Yes	Legal requirement.	• Incident response and event management framework. • Security incident reporting forms. • Security incident management reports.
5.29	Information security during disruption	Yes	To ensure at least the minimum required security controls are in place during BCP/DR scenarios.	• Business continuity plans. • Contingency policy.
5.30	ICT readiness for business continuity	Yes	To ensure the availability of the organization's information and other associated assets during disruption.	• Business continuity plans. • Contingency policy. • Record of testing and maintaining BCPs.

Control Reference (https://www.iso.org/)	Control Name (https://www.iso.org/)	Applicable Yes/No	Justification	If Yes, Give Reference Document
5.31	Legal, statutory, regulatory, and contractual requirements	Yes	Legal requirement.	<ul><li>List of applicable legislation.</li><li>Policy on compliance with legal requirements.</li></ul>
5.32	Intellectual property rights	Yes	Legal requirement.	<ul><li>IPR and data protection policy.</li><li>Software licenses.</li><li>Internal and external audit reports.</li></ul>
5.33	Protection of records	No	The organization is entirely paperless and securely stores all records with a secure cloud-based provider, making this control non-applicable.	N/A.
5.34	Privacy and protection of **Personally Identifiable Information** (PII)	Yes	Legal requirement as per the data protection law.	IPR and data protection policy.
5.35	Independent review of information security	Yes	To identify any significant deviations.	<ul><li>External audit reports.</li><li>Agreement with external auditors.</li></ul>

Control Reference (https://www.iso.org/)	Control Name (https://www.iso.org/)	Applicable Yes/No	Justification	If Yes, Give Reference Document
5.36	Compliance with policies, rules, and standards for information security	Yes	To validate compliance with information security policies and standards.	• Information security policy. • Information security audit procedure. • Internal and external audit reports. • Management review and follow-up reports.
5.37	Documented operating procedures	Yes	To ensure that operating procedures are available to every user and all users are aware of them.	Documented operating procedures.

Table 7.1 – Organizational controls

To sum up, organizational controls cover aspects such as the establishment of an ISMS, the management of information security roles and responsibilities, and compliance with relevant laws and regulations.

People controls

The people controls described in clause 6 of ISO 27002:2022 consist of eight controls. Information security is everyone's job, and these controls make sure that both internal and external users of an organization's systems know what they're supposed to do to keep sensitive data safe.

Controls include items such as job descriptions, background checks, training, awareness campaigns, disciplinary procedures, and the oversight of third-party vendors.

The people controls defined in ISO/IEC 27001 (and elaborated in ISO/IEC 27002) are enumerated in *Table 7.2*, the controls applicable to Titan Security Inc. are indicated, justifications for including or omitting controls are provided, and a corresponding reference document is also provided:

Control Reference (https://www.iso.org/)	Control Name (https://www.iso.org/)	Applicable Yes/No	Justification	If Yes, Give Reference Document
6	People controls			
6.1	Screening	Yes	All new employees are required to undergo background checks in accordance with the requirements of the company, client, as well as regulatory and legal regulations.	• Personnel security policy. • Screening procedure.
6.2	Terms and conditions of employment	Yes	The terms and conditions of employment are established in accordance with the human resources policies. The employees' responsibilities for information security are clearly established. This is also part of the contractual requirements.	• Personnel security policy. • The terms and conditions of employment are attached to the appointment letter.
6.3	Information security awareness, education, and training	Yes	To ensure that employees are aware of the policies and procedures and thereby enable them to perform their duties accordingly.	• Personnel security policy. • Training schedule. • Training material. • Training attendance records.
6.4	Disciplinary process	Yes	To ensure that the disciplinary process functions as a deterrent against malpractices, as a framework for taking appropriate actions in case of significant malpractice, and also to ensure that the organization is prepared legally.	• Personnel security policy. • Disciplinary procedure. • Record of disciplinary actions taken.

Control Reference (https://www.iso.org/)	Control Name (https://www.iso.org/)	Applicable Yes/No	Justification	If Yes, Give Reference Document
6.5	Responsibilities after termination or change of employment	Yes	To ensure that upon termination or change of employment, the employee is aware of the information security responsibilities that are still valid. This is to ensure confidentiality and prevent data leakage.	• Personnel security policy. • Termination or change of employment checklist. • Records of termination proceedings and correspondence.
6.6	Confidentiality or non-disclosure agreements	Yes	To ensure employees and relevant stakeholders comply with the protection of information.	Policy on compliance with legal requirements.
6.7	Remote working	Yes	To ensure information security is not compromised when connecting from outside the organization's network and to protect data accessed from alternate sites.	Mobile computing policy.
6.8	Information security event reporting	Yes	For effective reporting of information security events on the part of employees and stakeholders.	Incident response and event management framework.

Table 7.2 – People controls

Thus, people controls focus on controls related to the human aspects of information security, including personnel management, training and awareness, and the handling of sensitive information on the part of employees. It addresses areas such as the segregation of duties, background checks, and security education and training.

Physical controls

The purpose of the physical controls outlined in ISO 27002:2022 (clause 7) is to secure the storage, processing, and transmission facilities used to handle information assets. Controls aid businesses in avoiding security breaches, data loss, and interruptions in operations caused by the accidental or malicious destruction, theft, or misplacement of physical assets.

The physical controls section of the standard includes 14 controls that address the physical security aspects of an organization's information assets.

The physical controls defined in ISO/IEC 27001 (and elaborated in ISO/IEC 27002) are enumerated in *Table 7.3*, the controls applicable to Titan Security Inc. are indicated, justifications for including or omitting controls are provided, and a corresponding reference document is also provided:

Control Reference (https://www.iso.org/)	Control Name (https://www.iso.org/)	Applicable Yes/No	Justification	If Yes, Give Reference Document
7	**Physical controls**			
7.1	Physical security perimeters	Yes	All information processing facilities should be adequately secured at the perimeter to prevent unauthorized access to the facilities.	Physical and environmental security policy and procedure
7.2	Physical entry	Yes	Access control mechanisms should be in place to ensure only authorized personnel have access to the facilities	• Physical and environmental security policy and procedure • Visitors' register
7.3	Securing offices, rooms, and facilities	Yes	Physical security controls in accordance with the sensitivity of different zones will prevent unauthorized access to the facilities.	• Physical and environmental security policy and procedure • Swipe card system logs

Control Reference (https://www.iso.org/)	Control Name (https://www.iso.org/)	Applicable Yes/No	Justification	If Yes, Give Reference Document
7.4	Physical security monitoring	Yes	To protect against unauthorized physical access.	Physical and environmental security policy and procedure
7.5	Protecting against physical and environmental threats	Yes	To ensure resilience against risks associated with natural calamities, fires, and so on	• Physical and environmental security policy and procedure • Environmental system documents • Risk assessment
7.6	Working in secure areas	Yes	Access to secure areas such as data centers should only be for authorized personnel to ensure the protection of information.	Physical and environmental security policy and procedure
7.7	Clear desk and clear screen	Yes	To prevent the unauthorized disclosure of information through theft or unauthorized access.	Physical and environmental security policy and procedure
7.8	Equipment siting and protection	Yes	To ensure protection against equipment being damaged by environmental hazards and any unauthorized access.	Physical and environmental security policy and procedure
7.9	Security of assets off-premises	Yes	This is to prevent any unauthorized transfer of equipment, information, or software, thereby resulting in information leakage.	• Physical and environmental security policy and procedure • Equipment insurance documents

Control Reference (https://www.iso.org/)	Control Name (https://www.iso.org/)	Applicable Yes/No	Justification	If Yes, Give Reference Document
7.10	Storage media	Yes	To make sure that only authorized people can read, change, remove, or delete information from stored media.	Portable media policy
7.11	Supporting utilities	Yes	To prevent any loss/damage of information due to disruptions caused by power failures.	Physical and environmental security policy and procedure
7.12	Cabling security	Yes	Safeguarding cabling will prevent interception, interference, and damage, thereby preventing the loss of information.	• Physical and environmental security policy and procedure • Cabling security procedure and patch list
7.13	Equipment maintenance	Yes	Equipment maintenance shall be periodically performed as per the maintenance contract to prevent availability issues.	• Physical and environmental security policy and procedure • Daily equipment monitoring records and maintenance records • Equipment repair/servicing records • Equipment insurance documents

Control Reference (https://www.iso.org/)	Control Name (https://www.iso.org/)	Applicable Yes/No	Justification	If Yes, Give Reference Document
7.14	Secure disposal or reuse of equipment	Yes	To ensure that no sensitive data or licensed software resides on any of the equipment that is being disposed of, thereby reducing the chance of information leakage.	Physical and environmental security policy and procedure

Table 7.3 – Physical controls

To summarize, physical controls encompass those related to the physical protection of information assets and the secure management of physical environments. They include controls for physical access, the secure disposal of assets, protection against environmental threats, and the monitoring of physical security measures.

Technological controls

The technological controls outlined in ISO 27002:2022 are designed to protect an organization's technological assets, such as hardware, software, and network infrastructure, from security threats. It contains 34 controls, which are organized into several categories. Implementing these technological controls can help organizations protect their technological assets and prevent security incidents.

The technological controls defined in ISO/IEC 27001 (and elaborated in ISO/IEC 27002) are enumerated in *Table 7.4*, the controls applicable to Titan Security Inc. are indicated, justifications for including or omitting controls are provided, and a corresponding reference document is also provided:

Control Reference (https://www.iso.org/)	Control Name (https://www.iso.org/)	Applicable Yes/No	Justification	If Yes, give Reference Document
8	Technological controls			
8.1	User endpoint devices	Yes	To ensure the protection of information at user endpoint devices.	Mobile computing policy

Control Reference (https://www.iso.org/)	Control Name (https://www.iso.org/)	Applicable Yes/No	Justification	If Yes, give Reference Document
8.2	Privileged access rights	Yes	To ensure that privileges are granted as per the business requirements only to authorized users.	• Authorization records for the allocation of privileges • Logical access control policy
8.3	Information access restriction	Yes	To ensure that information and application systems are accessible only to authorized users.	• Digital access management policy • Digital access management matrix
8.4	Access to source code	Yes	To ensure the protection of intellectual property and prevent data leakage, access to the program source code should be restricted.	• Software development and testing policy • Audit logs of all accesses to program source libraries • Authorization records for personnel with access to the program source code • Change control records • Logical access control list of personnel authorized to access source code

Control Reference (https://www.iso.org/)	Control Name (https://www.iso.org/)	Applicable Yes/No	Justification	If Yes, give Reference Document
8.5	Secure authentication	Yes	To ensure that access to systems and applications is allowed only for authorized users using login credentials.	• Operating system security policy and procedure • Security message displayed when logging in • Logical access control policy
8.6	Capacity management	Yes	To ensure the availability of resources.	Policy on communications and operations management
8.7	Protection against malware	Yes	To prevent any unauthorized changes in the organization, business processes, and information processing facilities and systems that have not been authorized in the proper way, which could result in a loss of information and information assets.	• Policy and procedure for protection against malicious and mobile code • Logs of scans by the antivirus software

Control Reference (https://www.iso.org/)	Control Name (https://www.iso.org/)	Applicable Yes/No	Justification	If Yes, give Reference Document
8.8	Management of technical vulnerabilities	Yes	To ensure the protection of information assets and compliance with information security policies and standards.	• Software development and testing policy • Patch management procedure • Bug-fix procedure related to the OS, application, and network • Version control procedure • Policy on technical vulnerability management
8.9	Configuration management	Yes	To ensure the proper operation of all components.	Configuration management process document
8.10	Information deletion	Yes	To prevent the unnecessary exposure of sensitive information and conform to legal, statutory, regulatory, and contractual requirements for information deletion.	Asset management policy

Control Reference (https://www.iso.org/)	Control Name (https://www.iso.org/)	Applicable Yes/No	Justification	If Yes, give Reference Document
8.11	Data masking	Yes	To limit/prevent the exposure of sensitive data, including PII, and to comply with legal, statutory, regulatory, and contractual requirements.	Encryption policy
8.12	Data leakage prevention	Yes	To prevent the disclosure of sensitive information by individuals or systems.	Data protection policy
8.13	Information backup	Yes	To ensure all information is backed up and tested so that information is available in the event of a failure/disaster.	• Backup procedure (including a procedure for backup testing) • Record of backups taken • Logs of backup testing
8.14	Redundancy of information processing facilities	Yes	Redundancy of information processing infrastructure ensures availability.	Business continuity planning framework

Control Reference (https://www.iso.org/)	Control Name (https://www.iso.org/)	Applicable Yes/No	Justification	If Yes, give Reference Document
8.15	Logging	Yes	The implementation of this control is required to ensure that all activity that is relevant to the information security parameter defined for the asset is properly recorded and the data is retained for the period defined by the various business needs. This control helps in the investigation and evaluation of weaknesses and incidents reported.	• Event log policy and process • Information system monitoring policy
8.16	Monitoring activities	Yes	Monitoring anomalous behavior to prevent potential information security incidents.	Events and incident management process
8.17	Clock synchronization	Yes	Clock synchronization is implemented to ensure that there is uniformity of the timestamps across all devices to ensure that all implementations and changes can be made simultaneously. This is used for co-related logs and events during investigations.	Procedure for clock synchronization
8.18	Use of privileged utility programs	Yes	Access to the use of privileged utility programs (system utilities) is strictly controlled based on the needs of systems and applications.	Logical access control policy

Control Reference (https://www.iso.org/)	Control Name (https://www.iso.org/)	Applicable Yes/No	Justification	If Yes, give Reference Document
8.19	Installation of software on operational systems	No	The organization operates in a strictly controlled cloud environment, making it impossible for unauthorized software to be installed, rendering this control non-applicable.	N/A
8.20	Network security	Yes	To ensure that there is no unauthorized access to the organizational networks.	• Network security policy • Router and switch and firewall logs • Log monitoring records
8.21	Security of network services	Yes	To ensure that network security requirements are identified and documented for all services, thereby preventing unauthorized access to networks.	• Network security policy • Email policy and procedure

Control Reference (https://www.iso.org/)	Control Name (https://www.iso.org/)	Applicable Yes/No	Justification	If Yes, give Reference Document
8.22	Segregation of networks	Yes	Networks should be segregated according to the business requirements using VLANs and access should be controlled by access lists. This will ensure that there is no unauthorized access between groups of information services, systems, and users.	Network security policy
8.23	Web filtering	Yes	To prevent access to malicious content.	Network security policy
8.24	Use of cryptography	Yes	To ensure that critical information is protected using the appropriate encryption methods and key management.	Encryption policy
8.25	Secure development life cycle	Yes	To define and ensure the minimum required security when developing software and systems.	Secure software engineering development and testing process
8.26	Application security requirements	Yes	The encryption of information involved in application services in transit should be enacted as per the applicable requirements in order to ensure the protection of information.	Encryption policy

Control Reference (https://www.iso.org/)	Control Name (https://www.iso.org/)	Applicable Yes/No	Justification	If Yes, give Reference Document
8.27	Secure system architecture and engineering principles	Yes	To ensure security in application development and testing.	Secure software engineering development and testing process
8.28	Secure coding	Yes	To ensure that software is written securely, thereby reducing the number of potential information security vulnerabilities in the software.	• Coding guidelines • Code review process
8.29	Security testing in development and acceptance	Yes	To ensure that all development meets the minimum security requirements, thereby reducing the risk of information loss.	Secure software engineering development and testing process
8.30	Outsourced development	Yes	To ensure that all outsourced development follows the agreement and the required security requirements are met.	Vendor management process for software development, testing, and enhancement
8.31	Separation of development, test, and production environments	Yes	Segregation of duties and provision of access of defined roles and responsibilities to be adhered to to prevent any conflict or data loss.	Policy on segregation of responsibilities

Control Reference (https://www.iso.org/)	Control Name (https://www.iso.org/)	Applicable Yes/No	Justification	If Yes, give Reference Document
8.32	Change management	Yes	Existence of a change management process and team to ensure any changes in the process are routed through the team, assessed, and monitored.	• Operations management and change control policy and procedure • Change management records
8.33	Test information	No	The organization exclusively uses synthetic data for testing, thus rendering the protection of test data non-applicable.	N/A
8.34	Protection of information systems during audit testing	Yes	To protect information during audit testing and prevent non-repudiation.	• ISMS audit checklist • ISO standard

Table 7.4 – Technological controls

This category of controls relates to the technical aspects of information security. It includes controls for network security, system hardening, access controls, encryption, incident management, and security monitoring.

This case study on the selection of controls and preparing an SoA highlights the critical role of controls in information security management. The controls are categorized by ISO into organizational, people, physical, and technological categories. The SoA serves as a valuable tool for communicating the rationale behind control choices and demonstrating compliance with relevant standards. By continuously reviewing and updating the SoA, organizations can adapt to evolving risks and maintain a strong foundation for their ISMS.

In the next case study, we will see the incident management process at Titan Consulting Inc.

Case study #3 – information security incident management

Information security incident management is a critical process that involves detecting, responding to, and mitigating security incidents to minimize their impact. Let us investigate a case study on this.

An employee at Titan Consulting Inc. receives an email. Unfortunately, the employee clicks on the link in the email, which results in suspicious behavior on the employee's machine. The employee reports this to the information security team through the designated incident reporting channel.

Titan Consulting Inc. follows these steps for information security incident management:

1. Report the security incident.
2. An initial analysis and categorization of the incident is carried out by the information security team and communicated to the relevant stakeholders.
3. Identify the root cause and execute the correction(s) and corrective action(s).
4. Record the incident and add the learnings to the knowledge base.

The information security team does an analysis and observes that it was a phishing email. They request the IT team blocks the email sender from further communication. The IT team goes ahead and isolates the employee's machine from the organizational network and performs a thorough scanning of the machine. The findings include malware that infected the system because of the employee's action. The IT team takes steps to remove the malware and conducts analysis on the network to ensure that it hasn't spread across the network.

As the next step, the employee's login credentials are reset by the team to eliminate the risk of stolen credentials. Multi-factor authentication is enforced across the entire organization and threat intelligence tools are further leveraged to identify the extent of the damage because of the credential leakage.

A lack of awareness among employees is found to be the root cause of this incident.

The information security team then initiates an awareness program for employees across the organization on phishing and various other types of social engineering attacks. The training is made mandatory for all employees at the organization.

To conclude, Titan Consulting Inc. showcased a systematic incident response process consisting of preparation, detection, containment, eradication, and recovery. The proactive measures taken by the information security team, such as incident monitoring, timely response, and collaboration, played a crucial role in minimizing the impact of the incident. Furthermore, this case study highlights the significance of continuous improvement and learning from incidents to strengthen the organization's overall security posture.

Summary

In this chapter, we saw three case studies that depicted the concepts covered in the previous chapters. The first one was on ISMS implementation. The organization showcased a successful adoption of ISMS principles. By effectively aligning its security objectives with the business goals and engaging key stakeholders, it established a robust framework for managing information security risks. This case study highlighted the importance of a structured approach to ISMS implementation and the value it brings in safeguarding critical assets and maintaining the trust of stakeholders.

The second case study explained the selection of controls and the preparation of an SoA. It highlighted the significance of identifying and assessing the organization's unique risks and selecting appropriate controls to mitigate those risks effectively. The study emphasized the need for a well-documented SoA, which serves as a roadmap for implementing and maintaining the chosen controls, thereby enhancing the organization's overall security posture.

In the third case study of incident management, we gained insights into the importance of a structured approach to information security incident management. Through the organization's systematic steps of incident detection, analysis, containment, and recovery, they effectively minimized the impact of security incidents and protected their critical assets. It also emphasized the significance of ongoing training and continuous improvement to enhance incident response capabilities and ensure the organization's resilience against emerging threats.

In the next chapter, we will investigate the aspects of planning and executing audits.

Part 3:
How to Sustain – Monitoring and Measurement

Part 3, incorporating *Chapters 8 to 12*, addresses the practical elements of auditing an ISMS, providing essential information on auditing principles, execution, reporting, and the continuous improvement of the system. *Chapter 8* introduces the principles of auditing, explores the types of audits, and guides you through the planning process for effective audits. *Chapter 9* focuses on conducting an audit, explaining how to collect objective evidence, evaluate system effectiveness, and formulate findings and recommendations. In *Chapter 10*, the spotlight is on audit reporting, follow-up processes, and strategies for continual improvement, ensuring your ISMS stays effective and efficient. *Chapter 11* emphasizes the competencies and ethical conduct required of auditors, vital to uphold the integrity of the auditing process. Finally, *Chapter 12* concludes the book with real-world case studies that delve into audit planning, nonconformity reporting, and audit reporting. Altogether, these chapters equip you with the practical knowledge and tools to successfully audit an ISMS and drive its continual improvement.

This part has the following chapters:

- *Chapter 8, Audit Principles, Concepts, and Planning*
- *Chapter 9, Performing an Audit*
- *Chapter 10, Audit Reporting, Follow-Up, and Strategies for Continual Improvement*
- *Chapter 11, Auditor Competence and Evaluation*
- *Chapter 12, Case Studies – Audit Planning, Reporting Nonconformities, and Audit Reporting*

8

Audit Principles, Concepts, and Planning

The **International Organization for Standardization (ISO)** defines an audit as follows:

> *A systematic, independent and documented process for obtaining objective evidence and evaluating it objectively to determine the extent to which the audit criteria are fulfilled. (ISO 19011:2018 – Guidelines for Auditing Management Systems)*

In July 2018, in response to the demand for guidelines on integrated management system audits, the International Organization for Standardization published *ISO 19011:2018*, titled *Guidelines for Auditing Management Systems*.

It is a meta-standard that details the audit principles, planning, and execution processes for several types of management system audits including information security, environmental, and quality management audits.

One of the most important components of this guidance is to make sure that the objectives of the audit program are well-aligned with an entity's basic business objectives and that the needs and best interests of the clients and any other relevant stakeholders are considered (*ISO 19011, Clause 5.2*). In an audit, basically, three main things are investigated:

- What is documented by the company
- Evidence to support that the best practices are actually implemented
- The requirements defined by the ISO standard being audited against (for example, *ISO 27001*)

ISO 19011 lays down the principles to be followed by auditors and also how an audit program is to be planned and initiated. This chapter will walk you through the types of audits (such as first -, second -, and third - party), principles to be followed by an auditor while conducting an audit, and details about audit programs.

In this chapter, we will cover the following main topics:

- Different types of ISO audit
- Seven principles of auditing
- Additional guidance for auditors
- Audit programs

First, let us explore the different types of audits based on who takes up the role of an auditor.

Different types of ISO audit

The *ISO 19011* standard is intended to help businesses in auditing processes. In ISO standards, there are mainly two kinds of audits that can be carried out (see also *Figure 8.1*):

- **Internal audits (first - party)**: Internal audits are assessments and analyses made by businesses on their own management systems. These audits are carried out by the company. There are several resources available to guide businesses on how to carry out internal audits, but the *ISO 19011* standard is the most important of them. Internal audits are a crucial part of meeting the requirements of most standards for management systems. Internal audits for any management systems do not lead to ISO certifications.

- **External audits (second -party and third - party)**: When relevant stakeholders outside the business conduct an audit, it is called a second-party audit (such as those carried out by suppliers). Third-party audits are the most crucial ones and result in certifications, and are conducted by certification bodies.

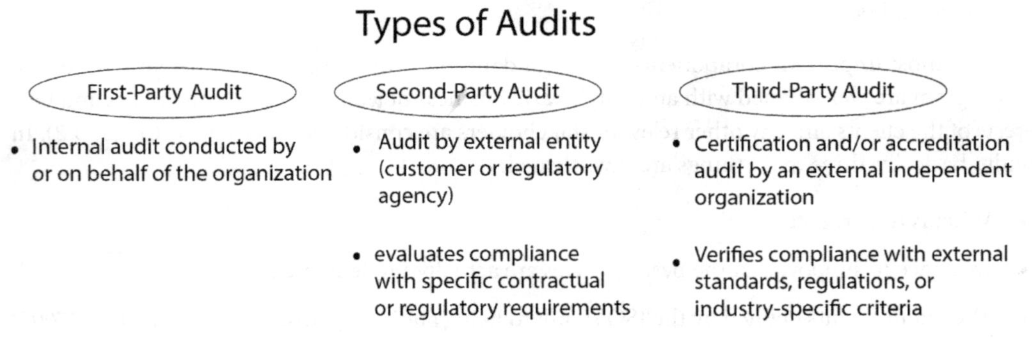

Types of Audits

First-Party Audit	Second-Party Audit	Third-Party Audit
• Internal audit conducted by or on behalf of the organization	• Audit by external entity (customer or regulatory agency)	• Certification and/or accreditation audit by an external independent organization
	• evaluates compliance with specific contractual or regulatory requirements	• Verifies compliance with external standards, regulations, or industry-specific criteria

Figure 8.1 – Types of audits

ISO 19011 is intended for use by audit teams of all sizes and types, from individual auditors to bigger teams suitable for comprehensive enterprise audits. Keep in mind that *ISO 19011* is a set of recommendations rather than a comprehensive set of instructions that must be followed in exact detail. To suit the needs and requirements of the audit program in context, the *ISO 19011* guidance should be adopted as necessary.

In *ISO/IEC 17021-1*, the specific standards for auditing management systems are described. These requirements are intended for use by trained lead auditors or registered organizations performing certification audits. However, *ISO 19011* can also be used as extra guidance for audits conducted by third parties.

Let's dig deeper into what each type of audit means.

First- party

A **first-party audit** is an internal audit.

Internal audits are carried out by the organization itself (or by someone on its behalf). Internal audits are often required by certain ISO standards that specify that they must be performed. Typically, these audits are undertaken for assessing conformance to the management system standard in context, evaluating effectiveness with respect to meeting the standard's requirements, and identifying improvement opportunities for the implementation of the management system.

First-party audits may also be conducted to prepare for third-party audits, but they will never lead to ISO certification.

Second- party

A **second-party audit** is carried out by relevant interested parties from outside of the business, such as customers or contractual organizations working on behalf of a client, or at the request of such parties.

For instance, a client and a vendor have a contract and exchange goods or services. The audit of the vendor performed by the client or on its behalf is a second-party audit for the vendor. Generally, second-party audits will be more formal than first-party audits, as they will affect relationships with customers and other relevant parties.

Third- party

Government agencies and organizations that provide certification are examples of those that perform **third-party audits**. These organizations are perceived as objective and free of bias when conducting audits. One of the fundamental characteristics of a third-party audit is the certifying organization's autonomy.

Third-party audits, at the request of a customer, are another way to prove your organization meets certain criteria. Obtaining ISO certification requires undergoing an independent audit. Registration, recognition, or licensing of various kinds may also result from independent audits. A fine or penalty could be issued for failing a third-party audit.

External audits encompass both second- and third-party audits.

The auditors' ability to execute audits effectively and fruitfully depends on their adherence to a set of fundamental auditing principles. Professionalism, independence, and diligence are all traits that should characterize an auditor's work. To be regarded as fair, an auditor's judgments must be the result of careful assessment of the auditee, and of course, the expertise of the auditor.

The next section covers the seven principles of auditing laid out by *ISO 19011* that an auditor must uphold to be professional in their conduct of an audit.

Seven principles of auditing

Auditing principles are needed for an audit to be an effective and reliable technique for supporting management systems and providing entities with chances for continual improvement. By adhering to these principles, audits can become effective and dependable tools in supporting management policies and controls. They provide essential information that organizations can act upon to enhance their performance. Following these principles is vital for generating relevant and comprehensive audit conclusions while enabling independent auditors to reach similar conclusions in similar situations.

Figure 8.2 depicts the seven principles laid out by *ISO 19011:2018, Clause 4*, followed by an explanation of each one.

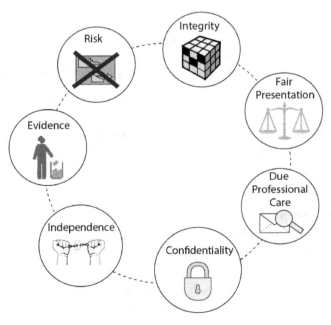

Figure 8.2 – Seven principles of auditing

Integrity – the foundation of professionalism

Auditors and individuals in charge of the audit program must conduct themselves ethically, honestly, and responsibly. They may only conduct audits if they have the relevant competence to do so. Being unprejudiced and unbiased and working in an impartial and fair manner is expected of them. When conducting an audit, an auditor must avoid being influenced by any factors that could compromise their objectivity.

Fair presentation – the responsibility to report truthfully and accurately

The audit processes carried out must be accurately and truthfully reflected in all audit findings, audit conclusions, and audit reports that the auditors present. All significant challenges encountered throughout the audit, as well as any unresolved differences between the audit team and the organization being audited, should be appropriately documented and disclosed. This should include any challenges or barriers encountered during the audit.

The communication should be timely, accurate, objective, and truthful. It should also be clear and thorough. This includes informing the client of the audit findings and delivering the audit report on schedule.

In conclusion, all conversations between the auditors and the parties under audit must be recorded, and any unresolved diverging opinions between the audit team and the auditee must be reported. All information reported must be accurate, timely, reasonable, clear, and comprehensive.

Due professional care – the application of diligence and judgment in auditing

Given the significance of the work they do and the trust the audit client and other interested parties have in them, auditors should take reasonable caution. Making informed decisions in all audit situations is crucial for completing one's task with the appropriate level of professionalism.

Confidentiality – security of information

Information gathered during an audit should be used and stored with discretion by auditors. This concept covers the secure handling of sensitive and confidential information by taking any additional precautions required, such as properly disposing of confidential documents after they are no longer needed and notifying the client right away if any leakage of confidential information is detected.

Due diligence must be used to guarantee that any information gathered is respected and securely stored. The auditor and audit client should not use this information improperly for their own gain or in any way that compromises the audit client's legitimate interests.

Independence – the basis for the impartiality of the audit and objectivity of the audit conclusions

When the auditor performs the audit operations, to the greatest extent of practicality, audits should by their very nature be impartial and objective. The auditor mustn't behave in a way that could give the impression that they are biased or have a conflict of interest.

Internal auditors are also expected to be impartial toward the person in charge of the function being audited. To make sure that audit findings and conclusions are founded entirely on facts obtained through audit evidence, auditors must maintain objectivity throughout the audit process.

In smaller organizations, auditors might not be entirely independent of the activity they are auditing. In such circumstances, the auditor and the entity should work together to investigate every avenue for preventing bias and fostering impartiality.

Evidence-based approach – the rational method for reaching reliable and reproducible audit results

A successful audit must have evidence, because it is the basis for results that are reasonable, reliable, and repeatable. It is essential that every piece of audit evidence that the auditors collect can be quickly verified. A sample of the information that is available is often used as evidence because an audit is normally conducted over a set period, as agreed upon with the audit client, and with limited resources.

In other words, audit evidence is compiled using a standardized procedure called audit sampling. A proper quantity of sampling should be used in an audit since it directly affects how much trust can be placed in the audit findings that the auditors present.

Audit conclusions must be based on facts rather than opinions or judgments if the auditor is to produce reliable and accurate audit results. Since this is closely related to the level of confidence that can be placed in the audit conclusions, an appropriate use of sampling should be put into practice.

Risk-based approach – the consideration of risks and opportunities

Planning, carrying out, and documenting an audit all require careful attention to risk management. A risk-based approach aids in guiding the audits more effectively toward issues that are crucial to the auditee's customers and the accomplishment of audit objectives.

The auditor must determine, based on the evidence at hand, which issues pose a serious risk to the audit goals. As a result, the audit will be carried out more successfully and quickly.

There are some extra points provided by *ISO 19011* as guidance for auditors that are explained in the next section.

Additional guidance for auditors

Annex A of ISO/IEC 19011:2018 provides additional guidance for auditors on different aspects explained in the following subsections.

Audit methods

An audit can be conducted using a number of methods including **on-site audit** and **remote audit**, with or without human interaction, in different combinations. On-site audits are those performed at the location of the auditee, and remote audits are those performed at a place different from the location of the auditee, regardless of the distance, using digital interactive communication means. Interactive audits involve the interaction of the auditee's representative(s) and the audit team. Non-interactive audits do not involve human interaction but involve interaction with equipment, facilities, and documents.

The objectives, scope, and criteria of the audit, as well as the audit's duration and location, all have a role in determining the audit techniques used. The audit process and its results can be improved by using distinct auditing techniques in combination with one another. The audit team leader is responsible for conducting the audit activities.

When multiple auditing organizations collaborate to conduct a joint audit of a common auditee, it is essential for the responsible individuals overseeing the respective audit programs to establish mutual agreement on the audit methods. This agreement should take into account the potential impact on resource allocation and audit planning. In cases where an auditee operates multiple management systems pertaining to different disciplines, it is possible to incorporate combined audits within the overall audit program.

Process approach to auditing

When a company has its management system audited, the processes and interactions inside the company are examined while considering one or more management system standards. It is important to view the organization's activities holistically as a system of interconnected processes. This is crucial for achieving stable and reliable audit outcomes. Moreover, the process approach in auditing is required by ISO on all of its management standards.

Professional judgment

Auditors should exercise professional judgment throughout the auditing process and focus on the desired outcome of the management system, as opposed to focusing on the precise requirements of each clause. They should use professional discretion to determine whether the clause's aim has been met. Also, some ISO management system standard clauses do not readily lend themselves to auditing in terms of direct comparison between a set of criteria and the content of a procedure or work instruction.

Performance results

The results of the management system and its performance are paramount. Though processes and their results are important, the intended result of the management system is what the auditors must prioritize.

The degree to which various management systems are integrated and their ability to deliver the desired outcomes is another factor to consider.

Verifying information

Auditors shall look for the following criteria when collecting information to assess whether it provides adequate, objective evidence to verify compliance with requirements:

- **Complete**: All the expected content is contained in the collected information
- **Correct**: The content is in accordance with other reliable sources such as standards and regulations
- **Consistent**: The documented information is inherently consistent with the connected resources
- **Current**: The information is up to date

Sampling

It is not possible to apply the audit process to the entire population in most cases. This is where sampling is implemented. When drawing conclusions about a population, auditors conduct audit sampling, which involves selecting a subset of data points rather than all of them. There are risks involved in sampling (that are to be communicated to the auditee at the beginning of the audit by the audit team), which include the following:

- Samples not being representative of the entire population
- Conclusions that differ from those that would have been reached had the complete population been investigated
- Risks depending on the variability within the sample and method chosen

Two main types of sampling are used in audits, which are as follows:

- **Judgment-based sampling**: This type is dependent on the expertise and experience of the auditing team. It takes into account prior audit experience within the audit's scope, the complexity of requirements, the challenges and interaction of organizational processes and management system elements, the evolution of technology, previously identified risks, and the output from monitoring of the management system. This method does not permit a statistical estimation of the impact of uncertainty on the audit's findings and conclusions.

- **Statistical sampling**: Variables such as the organization's setting, size, type, and complexity are considered. Also evaluated are the number of auditors, the rate of audits, and the duration of each audit. While selecting statistical sampling, the risk tolerance of the auditor is prioritized. This is referred to as the acceptable confidence level. A sampling risk of 5% equates to a 95% acceptable degree of confidence. This means that 5 out of 100 samples may not accurately reflect the actual results that would have been obtained if the entire population had been studied. A higher sampling risk suggests a lower level of confidence in the statistical inferences drawn from the sample. It signifies a higher possibility of the sampled results deviating from the actual results.

Auditing compliance within a management system

The audit team should evaluate whether the audited entity has efficient procedures for the following:

- Identifying its legislative and regulatory obligations as well as other commitments
- Monitoring its activities, products, and services to ensure compliance with these standards
- Doing a compliance evaluation

To verify compliance with applicable regulations, the audit team must assess whether the auditee addresses the following:

- A process for identifying changes in compliance standards and taking them into account as part of change management
- Those with the qualifications to oversee its compliance processes
- Adequate documentation of its current compliance status, if required by regulators and other interested parties
- Compliance standards being incorporated into the auditee organization's internal audit program
- Noncompliance instances
- Compliance performance being included in its management reviews

Auditing context

According to the requirements of management system standards, organizations must identify the context of their management system, which encompasses the needs and expectations of stakeholders as well as internal and external challenges. Audits should verify that relevant processes have been designed and are being utilized successfully. These processes offer the foundation for identifying the management system's scope and evolution. In the process of checking this, the auditors should look for objective evidence related to the methods used, the competency of individuals performing the process, the results of the process, and its application in the improvement of the management system scope and periodic reviews of context. The following two examples explain the interaction of organizational processes and what auditors can look for in them:

- **Incident management and continual improvement**: When a security incident occurs, the incident management process is activated to respond to and manage the incident. Information from this process (such as the type of incident, response, and outcome) is then fed into the continual improvement process to update and improve the organization's security measures. An auditor would look at the methods used in these processes, the effectiveness of the actions taken, and how the learnings were incorporated into improvements.

- **Employee onboarding and security awareness training**: When a new employee joins the organization, the HR process interacts with the security awareness process. The new recruit goes through security awareness training as part of the onboarding process. The effectiveness of this process interaction could be audited by checking the awareness level of the new employee, and whether they understand and follow the organization's security policies.

In both of these examples, auditors will be looking at the objective evidence associated with the method or process used, the individuals contributing to the process, the results of the process, how the results are applied, and how often the context is reviewed.

Auditing leadership and commitment

Most management system standards require the top management to demonstrate commitment by being accountable for the effectiveness of the management system and carrying out tasks expected of them, while delegating those that can be performed by others. In the audit, objective evidence must be gathered regarding the degree of involvement of top management in the decision-making processes related to the management systems and ensuring their effectiveness. This is accomplished by examining the policies, objectives, available resources, and proof of communications from senior leadership, as well as by interviewing the employees about the engagement of the leadership.

Auditors should interview top management to make sure they have a grasp of the ins and outs of the specific issues related to their management system. It's crucial for auditors to understand the organization's operating environment too, to ensure the management system delivers the desired outcomes.

Rather than just concentrating on leadership at the top level, auditors need to assess leadership and commitment across various management levels, as applicable, to ensure that every individual in top management is committed to the management system's success.

Auditing risks and opportunities

Auditing a company's approach to assessing its risks and opportunities should not be a standalone operation; rather, it should be implicitly integrated into the whole audit. The primary purpose of the operation is to assess the credibility of the risk and opportunity identification process, to determine whether the risks and opportunities have been appropriately identified, and to evaluate how the company tackles the identified risks and opportunities. In this regard, the audit steps should include the following:

- A review of the inputs used by the organization to determine its risks and opportunities (such as internal and external issues of the organization, interested parties, and potential sources of risk)

- Evaluation methodology, which can vary across disciplines and industries

Life cycle

A life cycle perspective evaluates the organization's ability to influence and manage the various phases of the product or service's life cycle such as raw material acquisition, design, production, transportation, use, end-of-life treatment, and disposal. The auditors must use their professional discretion to establish how the organization's strategy incorporates the life cycle perspective, considering aspects such as the duration of the product or service, the impact of the organization on the supply chain, the extent of the supply chain, and the technological intricacy of the product.

Audit of supply chain

The scope of a supply chain audit might vary based on the audit type, whether it is a complete management system audit, a single process audit, a product audit, or a configuration audit. A separate supplier audit program with adequate criteria for suppliers and external providers should be designed.

Preparing audit work documents

Audit work documents are the documents used to record information acquired during an audit. When preparing an audit work document, consider which audit record is to be created, the linked audit activity, the user of the document, and what information is needed to prepare the document. In combined audits, the work documents can be clustered for similar requirements of audit criteria, and the content from related checklists and questionnaires can be combined to avoid duplication. Audit work documents may be submitted in any format and must include every area of the management system within the scope of the audit.

Selecting sources of information

There are numerous information sources from which to choose evidence based on the extent and complexity of the audit, such as the following:

- Interviews with relevant personnel

- Observations of activities in the work environment

- Documented information such as policies, procedures, instructions, plans, objectives, licenses and permits, and so on

- Records such as minutes of meetings, audit reports, logs, inspection records, and results of measurements of any process outcomes

- Data analysis reports, summaries, and performance indicators

- Databases and company websites

- Simulation and modeling of any projects in context

- Reports from other sources such as customer feedback, news, and so on

Visiting the auditee's location

The visits as part of the audit must be planned while considering factors such as minimizing the interference with the auditee's work processes and the health and safety of the audit team.

A few points to consider while planning the visit are as follows:

- Auditors must obtain permission and ensure access to relevant areas in the auditee's location to be visited as per the audit scope (there can be NDAs between the auditing organization and auditee related to this).

- Auditors must be provided with adequate information on the health, safety, security, cultural norms, and working hours of the auditee organization. They should also confirm the need for any **personal protective equipment** (PPE).

- Auditors must check with the auditee the plans for using mobile devices and cameras, recording information, taking screenshots, making copies, and other similar actions while taking confidentiality and security concerns into account.

- The auditee must be informed of the audit objectives and scope unless it is an ad hoc audit.

On-site audit activities must make sure to avoid any unnecessary disturbance to the operational processes, schedule communications to minimize disruption, take care to avoid collecting personal information unless required by audit objectives/criteria, and take steps to decide on any change in the audit schedule in the occurrence of any impacting incident.

For virtual audit activities, remote communication means, devices, and software must be agreed upon between the audit team and the auditee. Audit evidence, including screenshots, must be collected only after receiving the auditee's permission and should be disposed of, regardless of the type of media, once the need for retention has lapsed.

Auditing virtual activities and locations

When a company operates or offers a service online, allowing individuals to accomplish tasks regardless of their physical locations, such as on the cloud, virtual audits are undertaken.

During a virtual audit, the standard audit process is followed but technology is utilized to verify objective evidence. In such audits, the auditor must possess the following:

- The technical skills to use the appropriate electronic equipment and other relevant technology during the auditing process

- Experience in facilitating virtual meetings to conduct remote audits efficiently

While performing such audits, the auditor should consider factors such as risks associated with remote audits. They should request permission in advance to capture screenshots or recordings, and ensure confidentiality and privacy during breaks by muting microphones or pausing cameras.

Conducting interviews

Interviews are a method of gathering information and should be conducted in a way that is appropriate for the circumstances and the person being interviewed, whether this happens face to face or through other forms of communication. While interviewing, the following aspects should be considered:

- Only individuals from appropriate levels and functions within the audit scope should be called for interviews.

- Interviews should be conducted during normal working hours and at the normal workplace.

- Attempts should be made to put the interviewee at ease prior to and during the interview.

- The reason for the interview should be explained to the interviewee. The interviews may start with giving the interviewer the chance to describe their work.

- Interviews can include open, closed, and guiding questions and appreciative inquiry, all of which should be carefully selected.

- Limited non-verbal communication is present in virtual settings, therefore focus should be more toward utilizing appropriate questioning techniques to gather objective evidence.

- The results of the interview should be reviewed with the interviewee.

- The interviewee should be thanked for their participation and cooperation.

Audit findings

While determining audit findings, consider inputs such as the follow-up of previous audit records and conclusions, the accuracy and sufficiency of the supporting objective evidence, the sample size, and the extent of realization of the audit plan. Recording conformities must include documentation of the applicable audit criteria, evidence to support conformity, and a declaration of conformity. Recording nonconformities must include the audit findings in addition to those documented for conformities.

In a combined audit, if the auditor identifies a finding related to a criterion in one management system, the possible impact on a similar criterion in other management systems should also be considered. The findings can be raised separately for each criterion or as a single finding combining references to multiple criteria. The auditee may receive advice from the auditor on how to approach such findings.

In the next section, let's see what an audit program is and what goes into devising one.

Audit program

An audit program is the master plan for conducting an audit or set of audits that are to be undertaken in a specific timeframe and for a specific purpose. For example, the purpose could be to certify the information security management system of a company against *ISO 27001*. It gives a direction for the proper execution of audits. The *ISO 19011:2018* standard offers instructions on how to manage audit program improvements in a systematic manner.

The objectives of the audit program should align with the policies and goals of the management system in addition to meeting regulatory and statutory requirements.

For third-party audits, the audit program must comprise an **initial audit** (Stage 1 – document review and Stage 2 – evaluating the implementation and effectiveness of the management system[s]), **surveillance audits** in the first and second years (after certification audits), and a **recertification audit** in the third year prior to the expiration of certification. The decision on certification or recertification marks the start of the three-year certification cycle. The scope and complexity of the client organization's management system, products, and processes, as well as the level of effectiveness of the management system that has been demonstrated and the findings of any prior audits, shall be taken into consideration when determining the audit program and any alterations that may be made later.

An audit program has the following advantages:

- It aids in making sure that all important areas are properly covered during the audit
- It helps in allocating tasks to the audit team members to provide assistance in accordance with their level of expertise and experience
- The audit team is given guidelines by an audit program, which also reduces the potential for misunderstanding
- As each member of the audit team is accountable for their own job, it is beneficial to establish clear lines of accountability within the audit team
- By comparing the amount of audit work that has already been accomplished to the amount that must still be done to effectively complete the audit, it is possible to gauge the progress of the task
- It provides evidence against a negligence claim
- If the audit is properly conducted, an audit program also functions as an audit record that may be used for future reference

If it appears necessary given the current situation, an auditor may decide to change the audit plan. The size of the entity, the industry in which it works, the efficacy of internal controls, the relevant regulations, and numerous other pertinent aspects will all have an impact on an audit program. As a result, an auditor creates an audit program in accordance with the audit's parameters. A few points to be considered in the process of framing an audit program are as follows:

- More audit program flexibility would be the solution to many problems. It's ought to be adaptable enough to take alterations into account and remain open to changes.

- Any flaws or errors in the audit program should be brought to the auditor's attention by the audit team that is responsible for conducting the audit.

- The audit staff should be encouraged to thoroughly investigate without being constrained by the audit program.

The audit program should take into account the auditee's organizational goals, pertinent internal and external concerns, interested parties' needs and expectations, information security regulations, and confidentiality demands. Areas with greater inherent risk and lower performance standards must receive priority when allocating resources. The program must identify resources that will allow the audits to be carried out properly and efficiently within the allotted timeframe and be managed by people with the necessary competency.

The audit program information is comprised of the elements shown in *Figure 8.3*. Until more thorough audit preparation is done, some of this information might not be available.

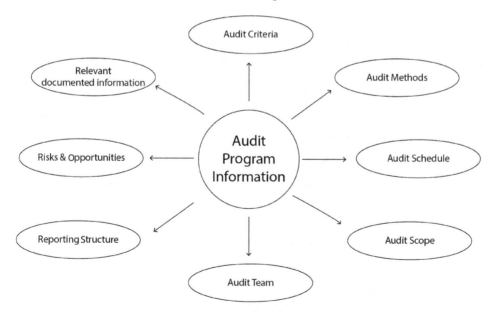

Figure 8.3 – Audit program information

Managing an audit program follows the **plan-do-check-act** (**PDCA**) cycle. *Figure 8.4* shows the process flow of the management of an audit program, in accordance with ISO 19011:

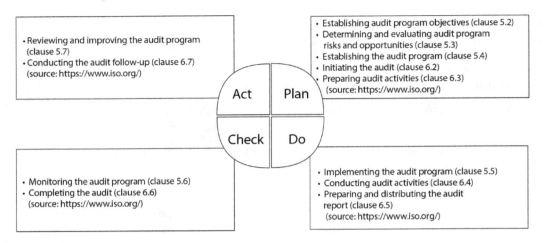

Figure 8.4 – Managing an audit program

Each process in the PDCA cycle, except the processes of performing an audit (*Clause 6*, which is explained in *Chapter 9*), is explained next.

Objectives

The audit program must have objectives established that would help in directing the planning and conducting of audits. The objectives are based on the following factors:

- Needs and expectations of interested parties

- Characteristics, requirements, and any modifications to processes, goods, services, and projects

- Requirements of the management system

- The need to assess external providers

- The auditee's performance level and maturity of the management system

Some examples of objectives are as follows:

- Identify opportunities for improvement of the management system

- Examine the auditee's capacity to identify its risks and opportunities

- Check whether the audited organization conforms to the relevant requirements such as statutory, regulatory, compliance, and management standards

Risks and opportunities

In the context of the auditee, there are risks and opportunities related to the audit program. Such risks must be identified by the audit program manager, who must also inform the audit client of them so that appropriate action can be taken.

Table 8.1 shows some of the risks and opportunities associated with an audit program:

Risk	Opportunity
Planning e.g. failure in scheduling audits	Enabling the conduct of several audits in a single visit
Resources e.g. allotted time or equipment not sufficient	Reducing travel time and distance to the site
Selection of audit team e.g. insufficient competency to conduct audits	Matching the audit team's competence to the level of competence needed
Communication e.g. ineffective communication processes/channels	Coordinating audit dates with the availability of the auditee's key personnel
Implementation e.g. inefficient audit coordination within the audit program	Continual training and development of auditors
Control of documented information e.g. failing to keep audit records to support the audit program's effectiveness	Leveraging technology, such as implementing audit management software
Monitoring, reviewing, and improving the audit program e.g. ineffective monitoring of audit program outcomes	Building a robust communication process to ensure effective and timely communication between auditors, auditees, and other stakeholders
The auditee's cooperation and availability of the evidence to be sampled	Periodic review and improvement of the audit program

Table 8.1 – Risks and opportunities associated with an audit program

Establishing the audit program

This process considers factors such as the roles, responsibilities, and competence of the personnel responsible for the audit program (for example, the audit program manager), the scope of the audit program, and the required resources.

The roles and responsibilities of an audit program manager include the following:

- Establishing the scope of the audit program
- Determining the internal and external issues, risks, and opportunities impacting the audit program, and measures to address them
- Ensuring audit teams have the appropriate competence
- Establishing procedures including processes for scheduling the audits, establishment of audit objectives and criteria, selection of an audit team, selection of audit methods, auditor evaluation, establishing internal and external communication processes, dispute handling, audit follow-up, and so on
- Provisioning the required resources
- Ensuring that the needed documented information is prepared and maintained
- Monitoring, reviewing, and improving the audit program
- Communicating the audit program to the relevant stakeholders

The audit program manager should have the following competencies:

- Knowledge of audit principles, methods, processes, the applicable management system standard, other relevant guidance documents, and statutory and regulatory requirements
- Understanding of the auditee and its internal and external context, needs and expectations of interested parties, business activities, products, services, processes, and so on
- Knowledge of risk, process, and project management
- Proficiency in the information and communication technologies needed
- Engagement in the continual improvement of the competence required for audit program management

Establishing the extent of an audit program

The audit program manager determines the extent of the audit program. Several factors such as those in the following list impact the extent of the audit program:

- Details of its context provided by the auditee
- The goal, duration, and scope of each audit, as well as the planned number of audits, reporting format, and audit follow-up
- Management system standards and any other related criteria
- The number, significance, complexity, similarity, and locations of the audited activities

- Elements affecting how well the management system works

- Applicable audit criteria, including prepared plans for the management system standards, legal and regulatory requirements, and other standards to which the auditee firm is committed

- Outcomes of prior internal and external audits, as well as management evaluations when applicable

- Outcomes of an evaluation of a previous audit program

- Concerns with language, culture, and society

- Interests of stakeholders

- Significant changes related to the auditee's context or operations, other internal or external events, and related risks and opportunities

- Accessibility of information and communication technology to facilitate auditing efforts

- Business possibilities and risks, as well as steps to address them

When establishing the audit program resources, the audit program manager must consider the following:

- The time and financial resources required to plan, carry out, oversee, and enhance auditing activities

- Audit methods

- Availability of auditors and technical specialists with the necessary skills to accomplish the goals of the audit program, both individually and collectively

- The scope of the audit program, as well as its potential threats

- Travel expenses, accommodation, and additional auditing requirements

- The effects of several time zones

- Accessibility of technologies for information and communication

- Accessibility of all necessary equipment, technology, and tools

- Accessibility of the necessary documentation

- Requirements for the facility, such as necessary equipment and security clearances (for example, background checks and PPE)

Implementation

Once the audit program has been formed and related resources have been identified, the audit program manager is required to undertake operational planning and coordination of all program operations, starting by establishing the following for each audit:

- Audit objectives
- Audit scope
- Audit criteria

All must be in line with the audit program's broad objectives. The audit's objectives outline what each individual audit is expected to identify and validate.

It is essential that the audit scope corresponds with the audit program's objectives. It considers things such as the locations, activities, functions, and processes to be covered, and the period of the audit. The following elements may be included in one or more of these: *statutory and regulatory requirements, management system requirements, appropriate policies, processes and procedures, performance criteria including objectives,* and so on.

In the case that the audit's objectives, scope, or criteria have to be altered in any way, the audit program should be amended as required and disclosed to any relevant parties in order to receive their consent, if necessary. The audit objectives, scope, and criteria must be consistent with the applicable audit programs for each discipline being audited when more than one discipline is to be audited concurrently.

Depending on the audit's declared objectives, scope, and criteria, the person(s) overseeing the audit program should select and identify the methods for successfully and efficiently conducting an audit. Both on-site and off-site audits, or a hybrid of the two, are viable options. The risks and benefits of using these techniques need to be weighed carefully. When two or more auditing organizations execute a joint audit of the same auditee, the individuals in charge of monitoring the multiple audit programs should agree on the audit processes and consider the consequences for resourcing and planning. If the auditee utilizes two or more management systems from other industries, a combined audit may be on the audit agenda.

The team leader and any technical expertise required for the audit should be decided by the person(s) administering the audit program. **Selecting audit team members** is done based on the competence required to complete the audit's objectives within the specified scope and should be taken into consideration when choosing the audit team. If there is only one auditor, the auditor should handle all pertinent leadership responsibilities for the audit team. The size and makeup of the audit team must consider the level of general expertise needed to accomplish the audit's objectives, the audit's complexity, whether it is a combined or joint audit, the auditing techniques employed, the audit's language and cultural characteristics, and other factors. Auditors-in-training must be supervised by the lead auditor. Before any changes are made to the team's composition, any competency or conflict of interest issues must be resolved between the audit team and the audit client.

The audit program manager should communicate the schedule well in advance to the audit team leader to ensure effective planning. The audit team leader must also be provided information on the audit objectives, scope, criteria, documented information, audit team composition, audit methods that will be employed, necessary resources, contact details of the auditee, locations, schedule, and any other additional information that enables the effective interaction of the auditor with the auditee. In the case of a joint audit, the organizations conducting the audits must reach an agreement on the responsibilities of each party.

The audit program manager is responsible for **managing audit program results** by evaluating the success of the audit program, reviewing and approving audit reports, reviewing the actions taken to address audit findings, disseminating audit reports to relevant interested parties, monitoring follow-ups, and sharing best practices with other departments.

The **audit records** can be any of the following three types:

- Related to the audit program, including the timeline, goals, scope, risks, opportunities, and so forth

- Related to specific audits, including audit plans, audit evidence and results, nonconformity reports, and so forth

- Related to the audit team, such as performance assessment and auditor competency

The personnel responsible for the audit program ensures that these records are generated, managed, and maintained. Also, any confidentiality requirements related to the records must be followed.

Monitoring the audit program

The audit program manager evaluates the success of the program by checking whether schedules are met, reviewing the performance of audit team members, considering the feedback from audit clients and other relevant parties, and inspecting the appropriateness and sufficiency of the documented information.

Audit findings, the level of effectiveness of the auditee's management system, conflicts of interest, legal and regulatory requirements, and the audit's scope all can contribute to deciding whether any modifications are required in the audit program.

Review and improvement

Learning from the review of the audit program should be associated with the program's improvement. The audit program manager examines the audit program's overall implementation, suggests opportunities for improvement, examines the ongoing professional development of auditors, and reports the audit program's findings.

The program review itself considers the outcomes and trends from its monitoring, compliance with processes, needs and expectations of relevant stakeholders, audit program records, new auditing and auditor evaluation methods, the effectiveness of addressing risks and opportunities of the audit program, confidentiality, and information security issues.

Thus, a well-defined audit program is key to the successful conduct of an audit, as is the role of the audit program manager.

Summary

ISO 19011 can be used by anyone who conducts or participates in audits or audit programs. It is intended for those who are tasked with overseeing audit program management and conducting audit participant evaluations. *ISO 19011:2018* is a useful resource for anyone who has been entrusted with enhancing an audit program. The standard outlines the seven auditing principles, how to manage an audit program, and methods for determining an auditor's level of competence. This chapter covered the theory, principles, and planning parts of conducting an audit, which may help you in better decision-making and planning for an audit in your organization.

In the next chapter, we will see how to conduct an *ISO 27001* audit. Again, the steps and processes involved are stated in the *ISO 19011* standard.

9
Performing an Audit

Auditing is important in maintaining trust and efficiency in a company's management system. An organization must foster a culture of continual improvement. They will be making worthwhile efforts for success, whether it be planning, scheduling, creating checklists, or carrying out or following up on audits, if they can pinpoint the areas to be audited. You can prevent inconsistencies between procedures and team member unavailability for the audit by creating a plan and schedule in advance. An **audit plan** assists your teams in organizing their operations and work properly. Scheduling audits will ensure that no procedure or area is overlooked.

Future audits should be announced to all departments as a courtesy so that they can be ready with the necessary documentation and other resources, such as evidence of the accomplished action plans, for the reviewer. Except in cases when there is a surprise audit of questionable behavior, your staff has every right to be informed about the organization's actions. By keeping them informed, you can help them understand how crucial audit procedures are to the ongoing success of your company. The audit checklist ensures that the proper questions are asked to the right people at the right time. To make sure that everyone is on the same page with regard to quality and brand reputation, a checklist frequently includes questions from audit team members about the auditee's process.

Immediately following the completion of the audit, you must document the audit findings. The audit findings should be adequate for a third-party reviewer to replicate and compare the results. After capturing the audit findings, you can conduct an analysis to determine which areas require immediate attention and which can be examined and addressed later.

The improvement areas indicated in the audit's findings will be studied to determine the underlying reason and an action plan will be developed appropriately. Finally, effectiveness evaluation and an audit follow-up are equally as vital as scheduling and conducting the audits since they allow you to determine the degree of an audit plan's effectiveness.

As you already know by now, ISO 19011 (named ISO 19011:2018 – Guidelines for auditing management systems; `https://www.iso.org/`) is a guide for organizations developing audit programs and conducting audits of existing **management systems**. In this chapter, we will explore the process of performing an audit as stated by ISO 19011.

In this chapter, we will cover the following main topics:

- An overview of the steps for performing an audit as per ISO 19011 guidelines
- Initiating the audit
- Preparing audit activities
- Conducting audit activities

In the following section, let's explore the steps for performing an audit as stated by ISO 19011.

An overview of the steps for performing an audit as per ISO 19011 guidelines

There are six main steps for conducting an audit, as listed here:

1. **Initiating the audit:**

   - Establishing initial contact with the auditee
   - Determining the feasibility of the audit

2. **Preparing the audit:**

   - Conducting a document review
   - Planning the audit
   - Task allocation to the audit team
   - Framing of documented information

3. **Conducting audit activities:**

   - Designating duties to guides and observers
   - Administering the opening meeting
   - Communication between the auditors and auditees and among the audit team members during the audit
   - Examining the availability and accessibility of audit information
   - Documented information examination while conducting the audit
   - Gathering and verifying information
   - Framing audit findings
   - Finalizing audit conclusions and conducting the closing meeting

4. **Preparing and distributing the audit report**

5. **Completing the audit**

6. **Conducting an audit follow-up**

Steps 1 to 3 are explained in this chapter. Steps 4, 5, and 6 come under audit reporting and follow-up and will be elaborated on in the next chapter.

In the next section, let's explore the first step.

Initiating the audit

The auditor must start the audit by getting in touch with the process owner and making sure the audit is possible. When doing an audit, it's better to make sure someone is there to present evidence than to try to catch them by surprise. It's also important to note that the audit team leader is the one who is ultimately responsible for how an audit is done.

Initiating an audit usually involves two steps, as mentioned here:

1. **Establishing initial contact with the auditee**: To establish contact with the auditee, the auditor and auditee decide on the communication channels. Both parties agree on the objectives, scope, methods, and composition of the audit team, which may include observers, guides, technical experts, or other roles. Any problems with the composition should be worked out at this point. The audit team intends to see relevant documents and records so that they can plan the audit. The laws and regulations that apply are considered. Since the audit process can involve dealing with confidential and critical data of the organization, the extent of disclosure of this is discussed and agreed upon by both parties. Also, any specific information related to health, safety, or access for the audit team must be communicated by the auditee.

2. **Determining the feasibility of the audit**: The feasibility of an audit is determined to decide whether audit objectives can be achieved. It is estimated by considering the appropriateness and sufficiency of information input for audit planning, the cooperation of the auditee, and the required time and resources.

In the case of adverse output, alternate plans are presented to the audit client.

Now, let's look at step 2 of performing an audit.

Preparing audit activities

The administration of the audit operations is covered in depth in the audit activities of ISO 19011. This methodical technique can assist in ensuring that your audits are efficient and reliable and strengthen the audit system. The individual steps in the process are detailed here:

1. **Performing the document review**: After the initiation process, the required documents must be reviewed. This helps in determining the extent of the system documentation for the audit and to analyze any gap which decides the audit plan in the following step. These documents include but are not limited to the following:

   - **Information Security Management System (ISMS)** scope and objectives

   - **Information Security (IS)** policy

   - ISMS risk register

   - **Statement of Applicability (SoA)**

   - Legal records

   - Monitoring and measurement records

   - Records of corrective actions

   - Management reviews

   The management system's documented information must be analyzed in order to comprehend the auditee's operations, plan for audit activities, and determine sufficiency in accordance with the audit criteria.

2. **Audit planning**: The document review contributes to the creation of an audit plan that specifies what will be audited, who will conduct the audit, when it will occur, and who will be audited. Here, you determine how the audit will be divided if more than one auditor will be employed, as well as how much time will be allocated to each audit process.

   A risk-based approach is adopted in planning, considering the information from the audit program and documented information from the auditee.

   To achieve audit objectives, it is necessary to plan out and organize the many steps involved. The audit client, audit team, and auditee should all be able to agree on the audit's conduct based on careful consideration of the risks posed by the audit itself. The depth and detail of the audit plan should correspond to the scope, complexity, and potential for failure to achieve audit objectives. The lead auditor considers factors such as the composition and competence of the audit team, sampling techniques, improvement opportunities for audit activities, risk of not achieving audit objectives due to ineffective planning, and risk to the auditee's operations due to the conduct of audit while devising the audit plan.

The strategies of planning can differ based on whether it is an initial or subsequent audit, an internal or external audit, and so on. The audit's goals, scope, criteria, and locations, the audit team's need to become familiar with the auditee's operations, the audit methods (on-site or remote), the audit team members' roles and responsibilities, and the audit's resource allocation should all be addressed at the planning stage. All aspects of the audit, including the auditee's representative, the audit language, logistics, communication arrangements, confidentiality factors, follow-up actions from prior audits, coordination requirements in the case of a joint or combined audit, and any other exclusive steps to be taken to achieve the audit objectives should be agreed upon by the audit team and the auditee during the planning phase. If there are issues with the audit plans, the auditee and the head of the audit team should discuss them.

3. **Assigning work to the audit team**: Larger audits may require the work to be delegated to multiple auditors, each of whom will audit multiple procedures. By having three auditors work for one day instead of one auditor working for three days, you can reduce how long an audit disrupts business operations.

    The leader of the audit team should collaborate with members of the audit team to determine which processes, activities, functions, or locations each team member will be responsible for auditing and empower them to make decisions regarding those areas. The leader of the audit team should convene meetings to assign responsibilities and discuss any necessary modifications. During an audit, work assignments may be altered to ensure that objectives are being met. All of this should be conducted with the auditor's independence, objectivity, competence, and resourcefulness, as well as the roles and responsibilities of the auditor, auditor-in-training, and technical experts, in mind.

4. **Preparing documented information**: The members of the audit team will review information pertinent to their audit tasks and generate documentation. Examples include *checklists*, *informational samples*, and *other audio-visual material*. The audit activities may also require additional resources. Documented information, often known as working documents, may consist of audit checklists, notes, recommendations, and other formats.

    In the audit working documents, the auditor will specify what it is that they want to check, what questions they want to ask, and what evidence they anticipate finding.

    The retention of this documented information is usually till the end of an audit or as specified in the audit program in alignment with the confidentiality and proprietary terms agreed with the auditee. An audit checklist can be particularly beneficial to the auditor as it assists with memory, helps in covering the required aspects, maintains continuity in audit, helps with time management, and can contribute to a well-written audit report.

Now that we have explored step 2 in detail, it's time to dive into step 3 of performing an audit.

Conducting audit activities

The steps in the process of conducting an audit are elaborated on as follows. There are no rules to adhering to this sequence but it can be tweaked according to the characteristics of specific audits:

1. **Assigning roles and responsibilities to guides and observers**: The presence of guides and/or observers is agreed upon by the audit team and auditee, following which they may accompany the audit team. They do not form part of the decisions during the audit and the audit team leader is the ultimate decision-maker who can at any point in time discontinue guides/observers if they interfere with audit activities. Guides usually help the audit team members arrange interviews with specific personnel, access specific locations in the organization, abide by the organizational rules of health, safety, or access, or make any general clarifications.

2. **Conducting the opening meeting**: The main purpose of the opening meeting is to discuss the audit plan and arrangements needed for the audit. It also provides the auditor an opportunity to set a friendly and collaborative tone for the audit and begin building a positive rapport with the auditees.

   This serves as a reminder to the people being audited that the audit will not come as a surprise and that its primary purpose is to confirm compliance rather than to identify areas for improvement. During the initial meeting, you will be able to make some adjustments to the audit schedule and will also be responsible for ensuring that everyone is aware of the scope and extent of this audit.

   There is a consensus arrived at regarding the scope; criteria and objectives; audit plan; audit team members and their roles; communication channels; language of the audit; confidentiality and integrity factors; health, safety, and access procedures; conduct of audit activities; and situations under which the audit may be terminated.

3. **Communicating during the audit**: It is recommended that lines of communication be kept open between all parties involved in the audit.

   If legislative and regulatory requirements necessitate **nonconformities** (**NCs**) to be reported, the audit team, auditee, audit client, and maybe other interested parties (for example, regulators) may need to formalize communication during the audit.

   The audit team should meet on a regular basis to share information, assess progress, and reassign responsibilities.

   Audit team leaders should regularly update the auditee and client on progress, important findings, and concerns. The auditor must immediately notify the auditee and audit client if the audit finds a substantial and urgent threat. Notify the audit team leader of any concerns regarding topics beyond the audit scope that should be communicated to the audit client and auditee.

   If the audit evidence indicates that the targets cannot be attained, the audit team leader is obligated to notify the audit client and auditee so that appropriate measures can be taken. This may mean changing the audit's goals or canceling it.

The auditee should be informed of any audit plan amendments approved by the audit program manager and audit client.

4. **Audit information availability and access**: The objectives, scope, criteria, duration, and location of an audit determine the audit methods. There can also be physical and virtual locations. It is very important to know when, where, and how to collect audit data. It doesn't matter where the information is created, where it is used, or where it is stored as these factors should not interfere with the need to have the data for auditing. The auditing processes should be chosen based on the previously mentioned criteria. The audit can take many different forms. Also, the circumstances of the audit may require a change in how the audit is done. Physical, virtual, and hybrid strategies can be adopted for the conduct of an audit depending on the circumstances and agreement between the parties.

5. **Document review**: By reviewing the documented information, the conformity of the system can be determined and information inputs to audit activities can be gathered.

   The review may be done alongside other audit operations without impacting the audit. The audit team leader should notify the audit program manager and auditee if enough written information cannot be provided within the audit plan's timeframe. Depending on the audit's aims and scope, the audit should continue or be put on hold until documented information issues are resolved.

6. **Collecting and verifying information**: The required and relevant information as part of the audit is collected by appropriate sampling.

   Only information that can be verified should be used as proof in an audit. For instance, you can ensure you comply with *control A 5.11, Return of Assets*, by double-checking your asset tracking log. Auditors should use their best judgment to decide how much weight to give evidence with a low level of verification.

   It is crucial to retain records of the audit evidence that supported findings. When conducting an audit, the team is responsible for addressing any issues, threats, or opportunities that are uncovered through the process of being objective. In addition to document review, information can also be collected via interviews and observations.

7. **Generating audit findings**: The auditor then has to come up with audit findings and make any audit conclusions before presenting them. Evidence gathered during an audit should be weighed against audit criteria to arrive at a conclusion. The results of an audit may show compliance or non-compliance with audit criteria. Each audit finding should include evidence of conformance and good practices, **Opportunities for Improvement (OFIs)**, and recommendations to the auditee, as required by the audit plan. NCs along with the evidence are recorded. NCs are usually graded as major or minor and are communicated to the auditee to ensure that they understand them and that the evidence is accurate. Any diverging opinions are to be resolved or recorded in the audit report. The audit team should review the findings at different stages during the audit.

Audit findings are graded into the following:

Conformity: Conformity is the fulfilment of a requirement. It is when the audited management system meets the requirements of the standard. There is factual evidence that the requirement has been met.

Nonconformity: NC is the non-fulfilment of a requirement. It is when the standard's requirement has not been met. There is factual evidence that the requirement has not been met. There will be a major nonconformance if there is no system in place to meet the ISO 27001 requirement or if that system has completely failed, for example, failing to adhere to regulatory requirements. Multiple minor NCs against a criterion will also be treated as a major NC.

Opportunities for improvement: Potential improvement areas for an organization. Here, the process or area has met the minimum requirement of the standard but can be improved.

Positives: Good practices in the organization.

If everything is determined to be complying, no corrective measures will be offered; however, if not, corrective actions must be carefully prepared. It is just as vital to emphasize best practices in a process as it is to address shortcomings. Some businesses also utilize audits to highlight OFIs, which the process owner can analyze and accept if appropriate.

8. **Audit conclusions and closing meeting:** The audit findings are presented during the closing meeting. This allows process owners to comprehend and ask clarifying questions. The audit team meets to discuss and agree on the audit findings and conclusions, frame recommendations, and decide on follow-ups in preparation for the closing meeting.

The conclusions must be factual and not based on assumptions. Audit conclusions address the following:

- The conformity of the management system with the audit criteria
- The effectiveness of the identification and addressing of risks by the auditee
- The implementation and maintenance of the management system by the auditee

After the audit is finished, the results and findings are presented in a closing meeting. The meeting is presided over by the audit team's leader and should include auditee management, audited function/process owners, the audit client, audit team members, and any other interested parties the audit team and auditee have mutually agreed upon.

The following are explained to the auditee in the closing meeting:

- The audit findings and conclusions
- The audit evidence was acquired on a sampling basis and is not indicative of the entire auditee process, and the auditors state
- How to address the audit findings as per agreed processes
- Post-audit activities, such as corrective actions

It is up to the audit team and the auditee to settle any discrepancies in opinion. If not resolved, then they need to be recorded. OFIs are also presented; however, are not mandatory to be implemented.

Summary

Everyone who conducts or participates in audits or audit programs can use ISO 19011. It is designed for individuals involved with supervising audit program administration and evaluating audit participants. ISO 19011 is a valuable resource for anyone tasked with improving an audit program. It is designed for those who administer audit programs or conduct audits (internal or external) of management systems.

There are six main steps established by ISO 19011 for the conduct of an audit, starting from initiating the audit to conducting follow-ups after the audit is over. Steps 1 through 3 were explained in this chapter. Audit initiation takes place first, between the auditee and the audit team. This is followed by preparing for the audit by reviewing the documents presented by the auditee, preparing the audit plan, assigning work to the audit team, and the preparation of documented information according to the work assigned. Conducting audit activities starts by defining the roles and responsibilities of guides and observers, followed by addressing the participants in the opening meeting. Periodic communication takes place between audit team members, while ensuring the accessibility and availability of information and collection of evidence takes place in parallel. Finally, the findings are generated and conclusions are finalized and presented to the auditee at the closing meeting.

In the next chapter, we will look at steps 4 to 6 in detail, that is, audit reporting and follow-up procedures and measures for continual improvement.

10
Audit Reporting, Follow-Up, and Strategies for Continual Improvement

In the previous chapter, we discussed the audit process, which involves steps such as initiating, preparing, and conducting the audit, reporting, completing the audit, and conducting a follow-up. The previous chapter explained the first three steps, and we will look at the remaining steps in this chapter. The closing meeting is the last activity of the steps in the audit process.

Following the closing meeting, a written record is created to deliver information in a formal manner, making it easy to follow up on the information. The process owner will obtain a higher value from the audit, which will enable changes to be made to the process. This will be accomplished by identifying not only the nonconforming parts of the process but also the good aspects and prospective improvement areas.

The step of following up is an essential component of the audit process, much like many other aspects of the ISO 27001 standard. If problems have been discovered and remedial steps have been taken, ensuring that the problem has been solved is an essential component of resolving it. If possibilities for improvement have been discovered and improvement initiatives have been completed because of the audit, then observing how much the process has improved is an excellent way to motivate more changes.

Plan, do, check, and act (**PDCA**) is an integral part of the ISO 27001 standard and forms the basis for continual improvement.

There are several different avenues available to enhance information security through the monitoring of many actions. To aid in the process of improvement, documenting the steps taken to meet ISMS objectives, risk treatment strategies, and management review decisions might be helpful.

We will cover the following main topics in the chapter:

- Audit reporting
- Completing the audit
- Conducting an audit follow-up
- Looking at the strategies for continual improvement

Let's start with the reporting process first. In the following section, we will see how audit reporting is done, what the components of an audit report are, how to report non-conformities, and how the report is distributed.

Looking at audit reporting

Audit reporting is how the outcome of an audit is conveyed to the client. The audit report shows all the audit information, its noncomplying items, observations, audit conclusions, and other comments. It is the lead auditor who is accountable for the accuracy and completeness of the report. The audit plan is where *what goes into the audit report* (the content) is decided.

In the audit report, the final results are formally documented and disseminated. In addition to giving a record of the audit's results, this provides everyone with a reference to the outcome of the audit. It is the client who becomes the owner of the report and decides its distribution. Ideally, an audit report should add value to the organization's **Information Security Management System (ISMS)**.

The audit report

Reporting audit findings is the responsibility of the lead auditor who must follow the audit plan. To give a full, correct, concise, and complete result of the audit, the audit report must contain the following:

- Audit objectives
- Details of the auditee and the functions and processes audited
- Details of the audit client
- Audit team and audit participant details
- Dates and locations of the audit
- Audit criteria
- Audit findings and observations
- Audit conclusions

- The extent of the fulfillment of audit criteria
- Unresolved differences of opinion between the auditee and the audit team
- A disclaimer regarding sampling adopted in the audit process

Conformities and nonconformities can turn up in an audit process. Reporting nonconformities requires much attention from the audit team, as it should be well understood by the auditee in order to take action.

Nonconformity report

Essentially, a **nonconformance/nonconformity report** (**NCR**) is a document that contains information about the requirement (the clauses in ISO 27001) that was not satisfied due to the **nonconformance/ nonconformity** (**NC**), how the NC occurred, and how to address the NC prior to deciding whether to take corrective action. The statement should be clear, unambiguous, concise, and supported by objective evidence. Make sure to include the audit criteria or the clause number from the ISO standard, against which the NC is marked. Ideally, an NC is graded as **major** or **minor** (as discussed in *Chapter 2*).

If there is at least one major NC, the audit will be halted and rescheduled. The certificate will be approved only after major NCs are fixed.

If there are any minor NCs, the auditee organization should submit a **corrective action plan** (**CAP**) to the auditor, and an ISO 27001 certificate will be issued upon the approval of the CAP. The corrective action implementation and effectiveness will be verified during the subsequent audits.

Rather than focusing just on meeting ISO standards, it is essential to remember that an NCR is meant to be used as a tool to identify and fix issues that keep the desired level of quality from being achieved. Auditees are required to agree, accept, and sign off on the report. In addition, details such as locations, auditee information, dates, and similar information should also be mentioned.

Always remember that timely reporting adds value!

> **Tips on report writing**
> - Report in simple terms
> - Avoid jargons
> - Use precise wording
> - Avoid abbreviations
> - Avoid the passive voice

Audit report distribution

The audit report should include a date, a review, and an approval. The audit report is forwarded to the customer by the lead auditor. The lead auditor is responsible for issuing the report within the time range specified in the audit plan. Any delays in the delivery of the report must be officially communicated to the client, and a new delivery date must be determined. The client is responsible for deciding how the audit report should be distributed to the various relevant interested parties. When it comes to the distribution of audit reports, the issue of confidentiality needs to be taken into consideration.

In the next section, we will see the activities that result in the formal completion of an audit.

Completing the audit

The term **completion** refers to finishing all the tasks that must be performed before the audit findings can be rendered and the audit procedures concluded.

The audit is complete after all audit activities have been done, or as otherwise agreed with the audit client. There may be unforeseen circumstances that prevent the audit from being performed in accordance with the audit plan, but this must be communicated and agreed upon with the client (including deciding on further steps).

When deciding whether to maintain specific records, all parties participating in the audit should reach a consensus on the matter, and such decisions should be made by taking into consideration the audit program as well as any applicable requirements.

Without the permission of the audit client and, where applicable, the auditee, the audit team and the individual(s) overseeing the audit program should not reveal any information collected during the audit or that is contained in the audit report to any other person. The audit client and auditee must be notified as soon as it is feasible of any audit document's contents.

After all the other processes have been finished, there are a few tasks that will inevitably take place as part of the conclusion of the audit process. These include the following:

- Obtaining written representations of what is proposed
- A concluding examination of the audit file
- Assessing discovered statements that are factually questionable

The audit's takeaways can help both the audit program and auditee anticipate and prepare for future risks and opportunities.

The following section explains the further steps after the conclusion of an audit. Often, this involves actions to be undertaken by the auditee before the certificate can be granted.

Conducting an audit follow-up

Typically, audit reports provide conclusions that can suggest actions based on the audit's objectives. *Clause 6.7* of *ISO 19011* (ISO 19011 is the standard that provides guidelines on determining audit objectives for an audit) includes identifying areas for improvement, evaluating conformance, evaluating capability, and evaluating effectiveness. Although a follow-up may seem a natural next step, it is not always the case.

If there is a third-party certification audit and the report indicates an action is required, it would be necessary to verify such activity before providing certification. If you perform an audit of a second-party provider, the customer will expect any NCs to be resolved. The same holds true for an internal audit; management will anticipate that the audit conclusions will be addressed. The audit program manager may sometimes believe that the audit process concludes with the audit report.

Depending on the audit's goals, the report's conclusion may suggest further steps. Determining the level of conformance, assessing capability, assessing effectiveness, and pinpointing areas for improvement are all audit objectives included in the *ISO 19011* standard, which gives guidelines on setting them. Verifying the completion of any corrective actions recommended by an audit of this type is required as a next step. All major NCs found in a third-party audit of a supplier must be resolved before certification can be issued. Likewise, the findings of an internal audit must be mitigated by the management.

While it is true that the primary objective of any audit is to determine the extent to which predetermined standards have been satisfied, the specific objectives of any given audit will vary greatly from one organization to the next. Audit results might be used to implement a strategy, confirm capabilities, or just for informational purposes, such as figuring out the risks of a new business.

As part of an audit or inspection, actions can be reviewed remotely instead of in person. Audit organizations may weigh the cost of a formal follow-up audit against its benefits.

Actions by an auditee for follow-up include the need for corrections, corrective, or improvement actions. In addition to these, the previous versions of **management system standards** (**MSS**) had *preventive action*. In order to align MSS with *Annex L*, preventive action is removed and replaced with areas such as *risk-based thinking* (*identifying potential areas that could impact the management system and where you are unsure of the outcome*).

Corrections is the ISO terminology for removing a discovered NC. It is like a quick fix. Corrective action, on the other hand, is an action taken to eliminate the source of the NC or undesired circumstance in order to prevent it from reoccurring.

The auditee normally decides and executes actions within a predetermined timeline. It is important to resolve findings as soon as possible, but it is even more critical to eradicate the source of the NC so that it does not reoccur.

ISO 19011 does not specify the necessity for a corrective action plan. Yet planning is an important component of problem-solving to ensure that future NCs from the same cause are avoided.

The benefit of establishing a corrective action plan is that management may review it to confirm that the solution will likely eradicate the cause(s) of the problem/NC. This strategy is more efficient than discovering during the follow-up audit that the solution will not resolve the issue. The quantity of planning will vary based on aspects such as the problem's simplicity or complexity, the authority level of the person(s) undertaking the planning, the competence of planners, and so on.

If an NC is critical, the solution must be expedited. According to the standard guidelines, the auditee should keep the person in charge of the audit program and the audit team informed of the status of these steps, as needed.

To track auditee actions, there are numerous ways and solutions available, including many electronic and software options. Audit program management should ensure that the software assumptions and data controls are consistent with audit program techniques and objectives when selecting software tracking programs.

Corrective improvement actions, in which a system is changed to eliminate the underlying problem, may be tried or piloted on a limited scale first. This procedure is recommended in order to avoid unanticipated consequences from a system-wide modification. Certain activities necessitate capital and other resources, which necessitates keeping the corrective action open for an extended length of time. Audit program managers can also plan a system-wide verification of a solution's implementation.

The auditing organization should then confirm that the activities were effective. The amount by which the intended activities and results are realized is characterized as effectiveness. This means that auditors must look into both the process and the outcomes. If the results (metrics) were accomplished and the process was capable and efficient, the corrective action would be effective. It is not rare for solutions to have unexpected repercussions, such as diminishing process efficiency or changing the output of a process to the point that it is no longer effective. The solution's test or pilot should disclose any issues.

This verification can be part of a subsequent audit or a one-time check. The follow-up can be done on-site or remotely through a virtual audit.

Audits are meant to benefit an organization and its business. All that's required for the audit's follow-up is for the auditee to respond to the findings, implement any necessary corrective actions, and check whether the measures have actually improved the situation.

The following list sums up the follow-up activities of an auditee:

- Understanding and examining the cause of the NC
- Identifying actions to eliminate the cause of the NC

- Selecting the most appropriate action and developing an action plan

- Implementing corrective actions

- Verification by the internal team at the organization

- Informing the audit team and conducting a follow-up

The following list sums up the follow-up activities of the auditor:

- Reviewing the corrective action plan

- Checking the corrective actions implemented via a remote/physical partial/full re-audit

- Confirmation of compliance after successful verification (recommendation for/continuation of certification)

In the next section, we will see the strategies for continual improvement, which is essential to keep up the effectiveness of the ISMS framework. It follows the PDCA process.

Looking at the strategies for continual improvement

To evaluate, test, review, and measure the success of the ISMS as a component of a larger business-led strategy, organizations that take improvement seriously need to conduct assessments, tests, and reviews.

Clause 10 of *ISO 27001* requires an organization to "*continually improve the suitability, adequacy and effectiveness of the information security management system*" (https://www.iso.org/). Documenting your continual improvement process is the most efficient approach to fulfilling this responsibility and complying with its requirements.

According to PDCA, each process that is carried out as a part of the management system needs to be planned, carried out, monitored, and improved. This method is an integral part of the standard and contributes to the continual improvement of ISMS (*Figure 10.1*). Continual improvement is the product of several procedures that make up the ISMS. The *ISO/IEC 27001* standard details the requirements that must be met for a company to create, implement, maintain, and constantly improve its ISMS.

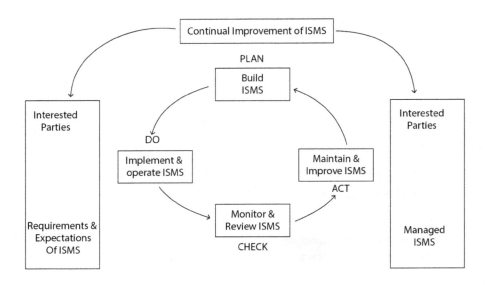

Figure 10.1 – The continual improvement cycle

From an ISMS implementation perspective, the *plan* stage (to build an ISMS) is followed by the *do* stage (implementing and operating the ISMS). The *check* stage is about monitoring and reviewing the implemented ISMS, and the *Act* stage is the measures taken to maintain and improve the ISMS based on the monitoring results. The interested parties/stakeholders give inputs in the process wherein they elaborate their requirements and expectations of the organization's ISMS. The final result is the implemented and managed ISMS by the organization.

Once the framework is implemented and measured, it is quite important to ensure continual improvement or improve the audit program's maturity. The objectives set during the framing of ISMS should satisfy the **Specific, Measurable, Attainable, Relevant, and Time-Based (SMART)** criteria. The set objectives must be periodically monitored and measured, the results of which should be considered to calibrate the set objectives whenever required. The opportunities for improvement need to be identified, and necessary controls to address the gaps should be implemented to catalyze the continual improvement of the management system.

Opportunities for improvement can come from a variety of sources, both internally and externally:

- Client requests or concerns
- Industry best practices
- Suggestions from inside an organization
- Potential risks

- Internal audits

- External audits

- Suggestions and observations from interested parties

The identification of an NC and the implementation of subsequent corrections and corrective actions can also give rise to continuous improvement programs. In this scenario, the lack of conformance might be interpreted as a chance to enhance a procedure, policy, or instrument.

Improvements can also come from a wide variety of different sources, and it is strongly recommended that the improvement process includes the documentation of these kinds of improvements.

The following areas can be monitored to identify opportunities for improvement:

- Annex A controls

- Policies

- Procedures

- ISMS objectives

One of the guiding principles behind every ISO standard is the idea that there should be ongoing progress. When you need to be able to demonstrate how you can continually enhance your ISMS, having an ISO 27001-certified management system offers a significant advantage.

Summary

To sum up, an audit report contains all the information about an audit, including its NC items, observations, findings, and other remarks.

The term **audit completion** refers to all of the essential activities that need to be finished before audit findings can be decided on and the audit processes can be brought to a close.

During an audit's follow-up phase, you'll deal with all of the findings, implement any necessary corrective and/or remedial activities, and make sure everything went according to plan.

As in the case of any business function, constantly trying to improve can only ensure the success of ISMS. As your ISMS matures or as you introduce a new process, you will be expected to revise your controls and rules to account for the resulting improvements. Your program will only be successful if you can pinpoint weak spots and fix them before they cause any problems.

In the next chapter, we will see the skill sets and competence requirements for an auditor.

11
Auditor Competence and Evaluation

There are three different aspects of auditor competence that are identified in the *ISO 19011* standard for management system auditing – *personal behavior*, *technical competence*, and *auditing competence*.

Auditors are required to have the relevant characteristics, knowledge, and abilities in each of these three domains. The key step to determining auditor competence is to determine which traits, knowledge, and abilities are required for each individual auditor for them to accomplish the goals that have been set for the audit and the audit program.

The required competency of an audit team can be determined based on a few factors, including the nature of the organization being audited, the nature and level of complexity of the audit that will be performed, the composition of the audit team, and any specific requirements that have been imposed by stakeholders.

It is not necessary for every auditor to possess the same level of expertise to conduct audits of all the components that make up an audit's scope.

It is essential to ensure that the combined expertise of the members of the audit team is commensurate with the aims and goals that have been established for the audit. To guarantee that this objective is met, it is up to the person who has been charged with the responsibility of putting together the audit team to take the necessary steps. In most cases, this is either the manager of the audit program or the head of the audit team.

The topics covered in this chapter are as follows:

- Personal conduct
- Knowledge and skills
- Auditor evaluation
- Maintaining and improving auditor competence

Let us investigate the first requirement of an auditor in the next section.

Personal conduct

Clause 4 of *ISO 19011* (`https://www.iso.org/`) describes the principles of auditing (explained in *Chapter 8*). Auditors are expected to exhibit attributes that enable them to abide by these principles. The individual behavior of an auditor should be professional during the conduct of an audit. ISO 19011 lists certain desired attributes for auditors, such as the following:

- **Ethical**: Auditors must base their findings on objective evidence and not falsify them for vested interests. Their reports should be truthful, have a defined objective, and be unbiased. Special care should be taken to refrain from unprofessional behavior, such as gossip or the disclosure of confidential information.

- **Open-minded**: Auditors must not let their own biases get in the way of assessing a process implementation by the auditee. Ultimately, the checks should be performed on whether the requirements are met or not and how effective they are in achieving the objectives set by an organization.

- **Diplomatic**: Auditors must try to maintain a neutral tone and avoid confrontation in their interviews with the auditee. They should focus more on open-ended questions and prompts that encourage auditees to explain more about their processes, rather than asking closed questions to get *yes* or *no* answers.

- **Observant**: Audits are a lot more than interviews. Auditors should watch out for nonverbal communication, the general setting of the work environment, safety and security protocols, placement of physical assets, and so on.

- **Perceptive**: The perception that you have is a direct product of your observations. An auditor needs to be able to interpret what a given observation implies, based on the inferences drawn from it. Auditors must have a solid understanding of a process for them to be able to make educated judgments about the conformity and efficacy of a system.

- **Versatile**: Auditing environments can be different based on locations, organizational culture, and various factors. An auditor should be ready to adapt to such differences while aiming to achieve set objectives. There can also be unforeseen incidents that disrupt the audit plan, in which case, the auditor should see how well the alternatives work. When necessary, the auditor must be flexible enough to revise a plan or make rearrangements in the audit team.

- **Tenacious**: For an auditor, it is important to get information. It might take a few attempts to ask the right question and get the required response. Auditors need to be persistent and not give up easily.

- **Decisive**: Deciding on findings and suggestions from observations requires effective decision-making on the auditor's side. An auditor is expected to reach conclusions in a timely manner, using logical reasoning and analysis.

- **Self-reliant**: An auditor must be able to act and make decisions independently based on their own abilities and skills, while simultaneously interacting effectively and professionally with others.

- **Acting with fortitude**: An auditor is always expected to act ethically and responsibly, even if this results in arguments or conflicts.

- **Open to improvement**: An auditor must also be a continual learner who is open to learning from new situations and challenges, aiming for effective audit results.

- **Culturally sensitive**: An auditor must respect and be considerate about the culture of the auditee and auditing organization, even though that may be far different from their own or something that they can't relate to. An audit may be carried out in multicultural environments, and thus, cultural sensitivities should be taken care of.

- **Collaborative**: An auditor should amicably collaborate and communicate with their fellow team members and the auditee's personnel.

This section covered the personal characteristics expected of a professional auditor. In the next section, we will see the domain knowledge and skills required to be a successful auditor.

Knowledge and skills

There are two types of intellectual skills that help auditors:

- Generic capability and sector-specific skills

- Competence in conducting audits

Let us see each in detail.

Generic knowledge and skills

The generic knowledge and skills of management system auditors are constituted by the following:

- **Knowledge of audit principles, processes, and methods**: This helps auditors in the effective conduct of an audit. With such competency, the auditor can plan the audit effectively as well as stick to the plan. They will also be able to implement a risk-based approach to auditing, as the auditor is aware of such approaches. The auditor will be able to plan and prioritize tasks efficiently to achieve the objectives. Communication as an important tool would be well understood by the auditor, and this would reflect in the audit interview process, enabling a more successful gathering of information and evidence. They will also verify the relevance and accuracy of collected information in order to only choose objective evidence. Employing sampling (selecting a subset of records from the population as a representation, for the purpose of the audit) during the audit process is crucial, and the auditor can recognize the consequences as well as convey them to the auditee. The processes audited can be mostly interrelated, and a knowledgeable auditor is aware of this. They also take care of the confidential information of the auditee revealed during the audit, as per their mutual confidentiality agreement.

- **Management system standards**: The audit scope and audit criteria are set based on the management system standards. In addition to these, there can be guidance or supporting documents. A standard forms the basis for conducting audits against requirements. An auditor who is knowledgeable of the standard can audit to check the fulfillment of its requirements by the auditee. The auditor can also understand and prioritize multiple standards where applicable.

- **An organization and its context**: Competency in an organizational context enables an auditor to verify the fulfillment of the needs and expectations of stakeholders that impact an audited management system. The auditor will also be able to assess the constitution of the organization in terms of size, governance, functions, cultural context, and so on. Knowing the general business and management of the auditee organization, including specific terminologies, helps in the smooth execution of the audit.

- **Statutory and regulatory requirements**: These help auditors understand the governing bodies impacting the auditee organization, the legal and regulatory requirements to be met by the auditee, the contracts, liability, and, in general, the basic terminologies used in these instances.

Discipline and sector-specific skills

When auditing management systems, the audit team collectively must have the discipline and sector-specific competencies to conduct the audit. The following are some of the elements that come under discipline and sector-specific skills:

- Practical knowledge of management system requirements and principles

- Knowledge of the relationship between the audited management system standards and the discipline/sector of the auditee

- Awareness of the sector-specific tools, techniques, methods, and processes to be implemented in the audit process, and also the need to evaluate the risks and opportunities of the auditee in their sector

Auditing multiple disciplines

According to the definition provided by ISO 19011, a combined audit is one that is performed by a single auditee on two or more management systems. The term *integrated audit* refers to a combined audit that is performed when multiple management systems have been incorporated into a single management system.

Audits of multiple disciplines that are carried out at the same time can either be categorized as combined audits or audits of integrated management systems that encompass multiple disciplines.

Integrated audits offer several advantages, including reduced certification costs, fewer audit interruptions, a simplified procedure, a decrease in documentation, and more uniform goals across many systems.

When auditing management systems that cover many disciplines, it is important for each member of the audit team to have a grasp of the interconnections and synergy that exist between the various management systems.

The heads of the audit teams need to have an understanding of the needs of each of the management system standards that are being audited, as well as an awareness of the limitations of their own expertise in each of the specialized areas.

Auditor and audit team leader competence

An audit team comprises the lead auditor and one or more auditors. All members are expected to have auditor competencies. The lead auditor has special responsibilities in addition to the normal auditors, which requires additional skill sets and knowledge. Let us investigate the competencies required for both categories.

Auditor competence

Auditors require a variety of skills, ranging from domain knowledge to interpersonal and communication skills. The path to achieving these competencies involves completing training programs for the generic skills, gaining practical experience in roles that exercise decision-making, problem-solving, and communicating with peers and interested parties, getting trained in relevant management system standards and conducting audits, and gaining experience under a lead auditor in the discipline. Completion of courses can be assessed based on participation and completing training modules, or by passing a designated score, based on the type of evaluation.

Audit team leader competence

The job of the lead auditor is an important one, since it bridges the gap between the representatives of departments and audit team members. ISO 19011 identifies *leadership* as one of the essential competencies of lead auditors, along with knowledge of *audit principles, methodologies, procedures, standards, information of auditee, and related requirements*. In addition, one of the most important skills that lead auditors need to have is the *ability to identify audit team members who are suited for the requirements of the audit team.* Building a productive team, on the other hand, may be difficult, just like any other endeavor that involves people. It is essential to bear in mind that the lead auditor might take further actions to acquire the competency of the audit team, such as providing new auditors with training and mentorship.

Even though having strong technical skills is essential for any audit team, that alone is not enough to ensure that they will produce accurate results. To achieve effectiveness in auditing, collaboration among the members of the audit team is one of the most important requirements, and the head of the audit team is the one responsible for enabling the team.

A few of the desired competencies of a lead auditor are as follows:

- Planning the audit and assigning tasks effectively to audit team members
- Communicating with top management of the auditee and getting the context of risk management at the auditee organization
- Collaboration among audit team members
- Managing the audit process at each stage by effectively communicating among audit team members, enabling the usage of resources, guiding the auditors-in-training, resolving conflicts, and so on
- Acting as the representative of the audit team in important communications with the audit program manager, audit client, and other stakeholders
- Guiding the team to reach audit conclusions
- Publishing the final audit report and handing it over to the audit client

The additional skills to be an audit team leader are acquired by working under the guidance of a different audit team leader(s) for a certain period.

Thus, the professional success of an auditor requires a combination of generic skills, domain-specific skills, and experience in auditing. The next section explains the importance of evaluating auditors and the criteria.

Auditor evaluation

The evaluation of auditors is an essential part of the ISO 19011 standard and one of its most important components. The evaluation of auditors is done with the intention of ensuring that those conducting audits have the requisite level of expertise, knowledge, and experience to carry out successful audits.

The evaluation of auditors can be carried out in several different ways, including self-assessment, a review by peers, and an evaluation by a higher level of management. The assessment must take into account not just the auditor's level of technical expertise but also their capacity to carry out the audit procedure in a manner that is compliant with the ISO 19011 standard. The standard also recommends that organizations establish a process for the selection and appointment of auditors and that they regularly review the performance of auditors to ensure that they continue to meet the necessary qualifications and requirements. This recommendation relates to the fact that organizations should establish a process for the selection and appointment of auditors.

Auditor evaluation has the overarching purpose of ensuring that audits are carried out in a way that is consistent, unbiased, and effective, and that the findings and recommendations of the audit are credible and correct.

Auditor evaluation criteria: The criteria for evaluation can be quantitative or qualitative. Quantitative factors include hours of auditing, education, hours of training, and other similar aspects. Qualitative factors include interpersonal behavior, performance in training, and so on.

A few of the evaluation methods are listed here:

- **Review of records**: This helps with the background check of the auditor. Verification of the records of training and professional credentials are examples of reviewing records.

- **Feedback**: An assessment of the performance of the auditor, conducted by surveys, questionnaires, a peer review, and so on.

- **Interview**: Personal interviews are conducted to verify skill sets, including communication skills and professional behavior, and to cross-check the credentials claimed.

- **Observation**: Witness audits and on-job monitoring are conducted to verify professional behavior and the ability and skills of an auditor.

- **Testing**: Assessments in the form of oral or written tests are conducted to evaluate the application of knowledge and skills.

- **Post-audit review**: After an audit, an auditor's performance can be assessed during the audit by interviewing the lead auditor or other auditors in the team, or by feedback from the auditee. Strengths, weaknesses, and areas of improvement are also identified.

While performing auditor evaluation, the results of the evaluation process are verified against the prescribed competencies of an auditor (*clause 7.2.3 of ISO 19011*). Any gaps identified are rectified with additional training or audit work hours. A further evaluation decides the results.

The next section describes the importance of maintaining auditors' competence.

Maintaining and improving auditor competence

Auditors build, maintain, and improve their level of competence through consistent participation in audits, as well as regular professional development activities. Competence should be something that auditors and leaders of audit teams consistently work to improve. Auditors should keep their auditing competence up to date by actively participating in management system audits regularly and engaging in ongoing professional development. This might be accomplished by the acquisition of further job experience, training, self-study, mentoring, attendance at meetings, seminars, and conferences, or participation in any number of other pertinent activities.

The audit program manager is responsible for aiding the continual evaluation of the auditor's performance in the audit team. Different types of training should be provided to upskill auditors as well as audit team leaders, based on the latest developments in auditing processes and technology enablement, changes in the needs of the auditor as well as the auditing organization, changes happening in the specific sector/industry, and upgradations in respective standards.

In a nutshell, quality auditing relies on having qualified auditors. To maintain competency, ongoing education and access to external resources are required. Both are necessary at the same time to carry out the most thorough audits.

Summary

As you can see, a variety of skills should be taken into consideration to have the right individual to carry out efficient audits of your organization's **information security management system (ISMS)**. Some of these skills are required while others are only desired. Even though it is conceivable for a person who lacks sufficient competence to conduct an audit of parts of ISMS, a professional auditor or audit team is your best bet to move your company closer to its goal of incorporating audits into its ongoing cycle of continuous improvement.

In the next chapter, we will see case studies based on audit planning, reporting NCs, and drafting the final audit report.

12

Case Studies – Audit Planning, Reporting Nonconformities, and Audit Reporting

This chapter aims to provide practical insights into the audit planning process, nonconformity reporting, and audit reporting within the context of ISO 27001 implementation. It strives to offer real-world examples that you can relate to and learn from. The following are the case studies presented in this chapter:

Case study 1 – audit planning

Case study 2 – reporting **Nonconformities (NCs)**

Case study 3 – audit reporting

These case studies revolve around a hypothetical organization named Titan Consulting Inc., a rapidly growing technology consulting firm operating in the IT industry with 50 employees. A total of eight employees work in the infosec domain.

The first case study on *audit planning* aims to help you understand the importance of thorough audit planning and how it contributes to the effectiveness of the audit process. It outlines the key considerations, steps, and best practices involved in developing an audit plan tailored to the specific needs of an organization such as Titan Consulting Inc.

The second case study on *reporting NCs* addresses the significance of reporting NCs identified during the audit process. It discusses the process of documenting and categorizing NCs, highlighting their implications, and providing guidance on remediation and follow-up actions.

The third case study on *audit reporting* delves into the elements of an effective audit report, including its structure, content, and communication strategies. It emphasizes the importance of clear and concise reporting to facilitate understanding, decision-making, and corrective action implementation.

The goal of the preceding case studies is to equip you with the knowledge and skills necessary to effectively undertake audit planning, report NCs, and communicate audit findings in the context of ISO 27001 implementation.

Case study 1 – audit planning

Audit planning is crucial in the ISO 27001 implementation process, as it ensures that the audit objectives are clearly defined, resources are allocated effectively, and potential risks and areas of focus are identified in advance, leading to a more efficient and comprehensive audit. It provides a structured approach to assessing the effectiveness of information security controls, identifying vulnerabilities, and determining the compliance level with the ISO 27001 standard, ultimately contributing to the continuous improvement of an organization's information security management system.

The following is the **audit plan** prepared for a third-party audit of Titan Consulting Inc. Details such as company information, the audit scope, team details, the audit activities, and who will be facing the audit (the client representative) are recorded in the audit plan:

Audit Plan	
Company:	Titan Consulting Inc.
Reference standard:	ISO 27001:2022
Site address:	Titan Consulting Inc., Seymour Street, London, UK
Certification scope:	Application development, HR, and infrastructure management within Titan Consulting Inc's premises. This is in accordance with the statement of applicability – version 2.0, dated Nov 1, 2022.
Audit date:	**Jan 5, 2023**
Audit type:	Stage two/initial audit
Audit language:	English
Lead auditor:	John Doe
Auditor(s)):	Ryan Smith
Technical/legal expert:	Matthew Right

Dec 15, 2022			
TIME	CLIENT PROCESSES	PROCESS OWNER	LOCATION
9:00 to 12:00 pm 2:00 to 5:00 pm	Audit planning	Lead auditor	Off-site

Jan 5, 2023			
TIME	**CLIENT PROCESSES**	**PROCESS OWNER**	**LOCATION**
9:00 to 9:30 am	Opening meeting	All	London
9:30 to 10:30 am	Physical security	In-charge/team	London
10:30 to 11:30 am	HR security	In-charge/team	London
11:30 to 12:30 am	IT network infrastructure	In-charge/team	London
12:30 to 1:30 pm	Lunch break		London
1:30 to 2:30 pm	Delivery (projects)	Project Manager/Project Lead/team	London
2:30 to 3:00 pm	Top management interaction	Central management	London
3:00 to 3:30 pm	Consolidation of audit results	Auditors	London
3:30 to 4 pm	Closing meeting	All	London

Jan 7, 2023			
TIME	CLIENT PROCESSES	PROCESS OWNER	LOCATION
9:00 to 12:00 pm 2:00 to 5:00 pm	Audit reporting	Lead auditor	Off-site

Date: December 29, 2022

Signature (lead auditor):_____

The case study provides an insight into the structure and format of a sample audit plan presented in the form of a table, which serves as a practical reference for creating your own audit plans. In this section, you have acquired guidance on how to effectively plan audits to assess and improve your organization's information security management system, in alignment with ISO 27001 requirements.

The next case study is about reporting NCs.

Case study 2 – reporting NCs

Reporting NCs during an audit is essential, as it enables organizations to identify and document deviations from established information security controls. It also provides valuable insights into areas that require corrective actions and improvements to maintain the integrity and effectiveness of their information security management system.

Major versus minor NC and OFIs

A major NC is a significant deviation or lapse in a system or process that either has led, or may potentially lead, to a failure to fulfil a requirement specified by a standard or regulation. This can include situations where a large part or all of a required system is either not implemented or ineffectively managed. Major NCs typically require immediate corrective action due to their severity, and they may significantly impact the quality, safety, or efficacy of the product or service, or pose a serious risk to the business or its customers.

An example of a major NC is an absence of risk assessment, where an organization has not performed a risk assessment to identify and evaluate information security risks. This is a major NC, as risk assessment is a key requirement of the standard.

A minor NC is a failure to meet a requirement that doesn't necessarily lead to a significant breakdown or failure of the system or process in question. It usually pertains to isolated or occasional lapses in the system, or when a single requirement of a standard is not completely met or implemented. While not immediately critical, minor NCs still require corrective action to prevent them from becoming major NCs in the future.

An example of a minor NC is incomplete records, where an organization has implemented an access control policy but does not consistently maintain records of access rights granted to users.

Opportunities for Improvement (OFIs) are suggestions or recommendations provided during an audit to enhance the existing systems, processes, or performance of an organization, going beyond merely achieving compliance with standards or regulations. These are not NCs but, rather, potential areas where changes could lead to increased efficiency, effectiveness, or customer satisfaction.

An example of an OFI is enhancing training, where an organization provides security awareness training to its employees and adds more real-world examples and regular updates on the latest cyber threats to make it more effective.

An NC report is a formal document that captures and communicates instances of non-compliance with established information security controls or requirements. It serves as a crucial tool for identifying, documenting, and addressing NCs, facilitating corrective actions and continuous improvement within an organization's information security management system.

During the audit, the following scenario was encountered by an auditor, which was decided to be an NC by the audit team. It was a major NC, and the format of reporting is depicted here.

During a third-party audit of Titan Consulting Inc., an examination of information security training records identified that there was no evidence of information security-specific training courses completed by all employees. The HR manager explained that formal training was imparted only to the personnel working specifically in the information security domain.

The scenario raises an NC based on clause 7.3 and control requirement A.6.3: "Personnel of the organization and relevant interested parties shall receive appropriate information security awareness, education and training and regular updates of the organization's information security policy, topic-specific policies, and procedures, as relevant for their job function" (source: https://www.iso.org/).

The following sample is an NC report that captures the specific NC in context:

ISMS AUDIT – NC REPORT	
Company under audit: **Titan Consulting Inc.**	NC number: **NC001**
Area under review: **A.6 People Controls**	ISO 27001:2022 clause number: **7.3**
Category: Minor	

Through effective reporting, organizations can address NCs, initiate corrective actions, and improve their overall security posture. The case study emphasizes the importance of a systematic and standardized approach to reporting NCs, enabling organizations to demonstrate their commitment to information security.

The next case study is about audit reporting.

Case study 3 – audit reporting

Audit reporting plays a crucial role in communicating the findings, observations, and recommendations derived from the audit process. It provides stakeholders with valuable insights into the effectiveness of controls, identifies areas for improvement, and facilitates informed decision-making to enhance the organization's information security practices.

The final audit report of Titan Consulting Inc. is shown here. The audit details are summarized first, followed by details of the findings and observations, including NCs.

Audit report		Report number: 1101	
Date of audit: Jan 5, 2023	Audit team: John Doe and Ryan Smith	Audit standard: ISO 27001:2022	
Areas audited (scope): Application development, HR, and infrastructure management within Titan Consulting Inc's premises. This is in accordance with the statement of applicability – version 2.0, dated Nov 1, 2022.	Auditee contact details: contact@titanconsulting.org Minor NCs raised – 1 Observations raised – 2	Major NCs raised – 0	

Summary of findings:

The organization has a defined ISMS policy.

Internal audit: Internal audits of the ISMS are scheduled and performed on a yearly basis. Observations of NCs are documented, and remedial measures are carried out if necessary. Prior audit findings are considered during the preparation phase of an internal audit.

The risk profile of the project is used as a planning basis for the audit.

A Q3–Q4 audit calendar for **Calendar Year (CY)** 2022 from Jun 5, 2022 to 10 Dec, 2022 was sampled, and the calendar covered the following:

- Project
- Auditee
- Month
- Status
- Audit date
- Closure date
- Auditor comment status
- Count of major NCs
- Count of minor NCs
- Count of observations

There was **one minor NC** – the absence of training imparted to all employees (stated in the NC report in detail).

The following are the identified **opportunities for improvement**:

- The effectiveness of a clear desk and clear screen can be improved
- The data retention schedule should consider all of the relevant legislation of the countries in which the business operates
- The use of employees' own personal computers could be subject to a formal **Bring Your Own Device (BYOD)** policy
- Controls for cloud security can be aligned with the ISO 27018 standard
- Attending ISMS awareness trainings can be used as a KPI for every employee

Statement of applicability

Statement of applicability – version 2.0, dated Nov 1, 2022

Conclusions/recommendations

Overall, the ISMS framework meets ISO 27001's requirements with strong management practices in place. There are opportunities for improvement, as stated previously in the OFI section, which will align the framework to support the information security posture of the organization.

The lead auditor's recommendation for ISO/IEC 27001:2022 is as follows:

The NC identified does not jeopardize the certification of the management system. Continued certification is therefore recommended, pending the acceptance of the corrective action plans(s) for the identified NC.

Auditor's signature

The case study presents an understanding of the key elements and structure of an audit report, including the scope, findings, and recommendations. It also demonstrates the importance of clear, concise, and objective reporting to effectively communicate the results of an audit and support informed decision-making for information security improvement.

Summary

This chapter introduced three case studies – an audit plan, reporting an NC, and an audit report, which are key artifacts in the audit process.

With this, we come to the end of the final chapter of ISO 27001 implementation and auditing. An effective information security framework implementation can enable an organization to be a security pioneer and stand out among its competitors. Implementing ISO 27001 management standards can be a guide for organizations to accomplish their security goals and have a robust security framework in place.

We hope you enjoyed reading the book. As we conclude, we hope you have equipped yourself with the knowledge and skills to establish and maintain an information security management system. There are numerous opportunities to apply this learning. Embrace the challenge of implementing ISO 27001 at your organization, leveraging the practical insights, external references, case studies, and best practices shared throughout this book. Use your acquired skills to drive a culture of security, protect sensitive information, mitigate risks, and enhance the overall resilience of your organization. Always remember that successful implementation is an ongoing process of continuous improvement. We wish you the best in your journey ahead!

Appendix – Terms and Definitions

Serial Number	Term	Definition
1	Access control	Grading access to assets is restricted, based on business and security considerations on a need-to-know basis.
2	Analytical model	The algorithm or computation that combines several different decision criteria with a number of different base metrics.
3	Attack	An attempt to damage, expose, or change an asset in any way, steal it, or utilize it without authorization.
4	Attribute	A trait or feature of an object that may be quantified or qualitatively identified by human or automated means.
5	Audit	An objective procedure to review audit data and evaluate whether or not the audit criteria have been met through systematic, independent, and documented means.
6	Audit scope	The audit's scope and bounds.
7	Authentication	Giving assurance that an entity's claimed characteristic is correct.
8	Authenticity	The property of an entity that assures it is what it claims to be.
9	Availability	The ability to be accessed and used by an authorized entity on demand.
10	Base measure	An attribute-based measurement that is used to quantify something.
11	Competence	The ability to put skills and knowledge to work in order to attain desired outcomes.
12	Confidentiality	The property of not making information available or disclosing it to unauthorized people, organizations, or procedures.
13	Conformity	The realization of an objective.
14	Consequence	An event's outcome that has an impact on goals.

Serial Number	Term	Definition
15	Continual improvement	Continual action to improve performance.
16	Control	A risk-modification measure.
17	Control objective	A statement that describes what will be accomplished as a result of putting controls in place.
18	Correction	The action taken to mitigate an identified nonconformity.
19	Corrective action	The action taken to eliminate the recurrence of a nonconformity in the future.
20	Data	Values assigned to base measurements, derived measures, and/or indicators.
21	Decision criteria	The thresholds, targets, or patterns used to decide whether or not action or further study is required, as well as to express the level of confidence in a particular outcome.
22	Derived measure	The measure that is derived from the values of two or more base measures.
23	Documented information	The information that an organization must control and maintain, as well as the medium on which it is stored.
24	Effectiveness	The degree to which planned activities are carried out and expected outcomes are obtained.
25	Event	The occurrence or change of a particular set of circumstances.
26	Executive management	A person or group of persons to whom the governing body has assigned responsibility for implementing plans and policies in order to achieve the organization's mission.
27	External context	The external environment in which an organization aspires to accomplish its goals.
28	Governance of information security	The system that directs and controls an organization's information security efforts.
29	Governing body	A person or group of persons who are responsible for an organization's performance and compliance.
30	Indicator	A metric that gives an estimation or evaluation of specified attributes obtained from an analytical model, in relation to given information requirements.

Serial Number	Term	Definition
31	Information need	The understanding required to effectively manage objectives, goals, risks, and difficulties.
32	Information processing facilities	Any data processing system, service, or infrastructure, as well as the physical location where it is located.
33	Information security	Maintaining data's confidentiality, integrity, and accessibility.
34	Information security continuity	The methods and procedures in place to ensure that information security activities continue to run smoothly.
35	Information security event	A situation that was previously unknown but may have significant security implications, or a discovered occurrence of a system, service, or network state that suggests a potential violation of information security policy or control failure.
36	Information security incident	A single or series of unwanted or unexpected information security events that have a significant probability of compromising business operations and threatening information security.
37	Information security incident management	Information security incident detection, reporting, assessment, response, management, and learning processes.
38	Information sharing community	Participants in an information-sharing agreement.
39	Information systems	A collection of data-handling programs, services, or other IT assets.
40	Integrity	Attributes of precision and comprehensiveness.
41	Interested party	A person or organization that has the ability to influence, is impacted by, or feels they are impacted by a choice or an action.
42	Internal context	The internal factors in which an organization aims to fulfill its objectives.
43	ISMS project	An organization's organized activities to implement an ISMS.
44	Level of risk	The consequences and likelihood of something happening.

Serial Number	Term	Definition
45	Likelihood	The chances that something will happen.
46	Management system	Interconnected or interacting organizational components that collaborate to set policies and goals, implementing the processes required to achieve those goals.
47	Measure	A variable that is assigned a measurement-based value.
48	Measurement	The method used to arrive at a value.
49	Measurement function	The combination of two or more base measures using an algorithm or calculation.
50	Measurement method	A set of steps followed in order to arrive at an attribute's numerical value on a specified scale.
51	Measurement results	One or more indications, as well as their interpretations, that address an information requirement.
52	Monitoring	Determining a system's, process's, or activity's state.
53	Nonconformity	Noncompliance with a requirement.
54	Non-repudiation	The capacity to prove the existence of a stated event or action, as well as the entities responsible for it.
55	Object	Item specified by the estimate of its attributes.
56	Objective	The desired outcome.
57	Organization	A person or group of people with their own functions, responsibilities, authorities, and relationships to achieve their goals.
58	Outsource	An arrangement in which an outside organization performs a portion of an internal organization's function or process.
59	Performance	A measurable outcome.
60	Policy	The stated goals and objectives of an organization's leadership.
61	Process	A set of linked or interacting operations that translates input to output.
62	Reliability	The quality of being consistent in your intentions and the results you expect.
63	Requirement	A declared, commonly implied, or legally mandated requirement.
64	Residual risk	Risk present after risk treatment.
65	Review	The determination of a subject matter's acceptability, adequacy, and effectiveness in achieving stated objectives.

Serial Number	Term	Definition
66	Review object	Review of a specific item.
67	Review objective	A statement stating what a review is supposed to accomplish.
68	Risk	The impact of uncertainty on goals.
69	Risk acceptance	A decision to take a given risk based on sound reasoning.
70	Risk analysis	Understanding the nature of risk and determining the amount of risk.
71	Risk assessment	Identifying, analyzing, and evaluating potential risks.
72	Risk communication and consultation	The ongoing and iterative ways in which a corporation engages in risk management dialogue with its customers and partners.
73	Risk criteria	A set of criteria through which the significance of risk is assessed.
74	Risk evaluation	Determining whether or not a risk and/or its magnitude can be tolerated or controlled by comparing the findings of risk analysis with risk criteria.
75	Risk identification	The procedure to discover, identify, and characterize possible risks.
76	Risk management	Comprehensive risk management activities that advise and govern an organization.
77	Risk management process	Applying policy and procedure to the activities of communication and consultation to set a framework and detect, analyze, evaluate, treat, monitor, and review risk systematically.
78	Risk owner	A person or organization tasked with managing risk.
79	Risk treatment	A procedure to reduce or eliminate risk.
80	Scale	A sequence of continuous or discrete values, or a collection of categories to which the attribute is mapped.
81	Security implementation standard	A document outlining the officially approved methods of achieving security.
82	Stakeholder	A person or group that has the potential to influence, is influenced by, or perceives themselves to be influenced by a choice or activity.
83	Threat	Something that might potentially cause problems for an organization or information systems.

Serial Number	Term	Definition
84	Top management	An individual or group in charge of an organization's top levels of management and control.
85	Trusted information communication entity	A self-governing organization that promotes communication within a sharing community.
86	Unit of measurement	A specific amount of a given quantity that may be expressed numerically.
87	Validation	The existence of objective evidence confirming that the requirements for a certain planned use or implementation have been satisfied.
88	Verification	Verifying compliance with a set of conditions by providing objective evidence.
89	Vulnerability	A flaw in an asset or control that one or more threats can exploit.

Index

Packtpub.com

Subscribe to our online digital library for full access to over 7,000 books and videos, as well as industry leading tools to help you plan your personal development and advance your career. For more information, please visit our website.

Why subscribe?

- Spend less time learning and more time coding with practical eBooks and Videos from over 4,000 industry professionals

- Improve your learning with Skill Plans built especially for you

- Get a free eBook or video every month

- Fully searchable for easy access to vital information

- Copy and paste, print, and bookmark content

Did you know that Packt offers eBook versions of every book published, with PDF and ePub files available? You can upgrade to the eBook version at packtpub.com and as a print book customer, you are entitled to a discount on the eBook copy. Get in touch with us at customercare@packtpub.com for more details.

At www.packtpub.com, you can also read a collection of free technical articles, sign up for a range of free newsletters, and receive exclusive discounts and offers on Packt books and eBooks.

Other Books You May Enjoy

If you enjoyed this book, you may be interested in these other books by Packt:

Cybersecurity Blue Team Strategies

Kunal Sehgal, Nikolaos Thymianis

ISBN: 978-1-80107-247-2

- Understand blue team operations and its role in safeguarding businesses
- Explore everyday blue team functions and tools used by them
- Become acquainted with risk assessment and management from a blue team perspective
- Discover the making of effective defense strategies and their operations
- Find out what makes a good governance program
- Become familiar with preventive and detective controls for minimizing risk

Digital Forensics with Kali Linux - Third Edition

Shiva V. N. Parasram

ISBN: 978-1-83763-515-3

- Install Kali Linux on Raspberry Pi 4 and various other platforms
- Run Windows applications in Kali Linux using Windows Emulator as Wine
- Recognize the importance of RAM, file systems, data, and cache in DFIR
- Perform file recovery, data carving, and extraction using Magic Rescue
- Get to grips with the latest Volatility 3 framework and analyze the memory dump
- Explore the various ransomware types and discover artifacts for DFIR investigation
- Perform full DFIR automated analysis with Autopsy 4
- Become familiar with network forensic analysis tools (NFATs)

Packt is searching for authors like you

If you're interested in becoming an author for Packt, please visit `authors.packtpub.com` and apply today. We have worked with thousands of developers and tech professionals, just like you, to help them share their insight with the global tech community. You can make a general application, apply for a specific hot topic that we are recruiting an author for, or submit your own idea.

Share Your Thoughts

Now you've finished *Mastering Information Security Compliance Management*, we'd love to hear your thoughts! Scan the QR code below to go straight to the Amazon review page for this book and share your feedback or leave a review on the site that you purchased it from.

`https://packt.link/r/1803231173`

Your review is important to us and the tech community and will help us make sure we're delivering excellent quality content.

Download a free PDF copy of this book

Thanks for purchasing this book!

Do you like to read on the go but are unable to carry your print books everywhere?

Is your eBook purchase not compatible with the device of your choice?

Don't worry, now with every Packt book you get a DRM-free PDF version of that book at no cost.

Read anywhere, any place, on any device. Search, copy, and paste code from your favorite technical books directly into your application.

The perks don't stop there, you can get exclusive access to discounts, newsletters, and great free content in your inbox daily

Follow these simple steps to get the benefits:

1. Scan the QR code or visit the link below

https://packt.link/free-ebook/9781803231174

2. Submit your proof of purchase
3. That's it! We'll send your free PDF and other benefits to your email directly

www.ingramcontent.com/pod-product-compliance
Lightning Source LLC
Chambersburg PA
CBHW060548060326
40690CB00017B/3641